Fundamentals
of
Electrical Control

Fundamentals
of
Electrical Control

by

Clarence A. Phipps

Published by
THE FAIRMONT PRESS, INC.
700 Indian Trail
Lilburn, GA 30247

Library of Congress Cataloging-in-Publication Data

Phipps, Clarence A., 1922-
   Fundamentals of electrical control / by Clarence A. Phipps.
   p. cm.
   Includes index.
   ISBN 0-88173-217-6
   1. Electronic control. 2. Logic design. 3. Programmable logic devices.
I. Title.
TK7881.2.P55   1995   629.8'043--dc20   95-19042
                                                         CIP

Fundamentals Of Electrical Control By Clarence A. Phipps.

Published by The Fairmont Press, Inc.
700 Indian Trail
Lilburn, GA 30247

Printed in the United States of America

10 9 8 7 6 5 4 3 2 1

ISBN 0-88173-217-6   FP

ISBN 0-13-504846-X   PH

While every effort is made to provide dependable information, the publisher, authors, and
editors cannot be held responsible for any errors or omissions.

Distributed by Prentice Hall PTR
Prentice-Hall, Inc.
A Simon & Schuster Company
Upper Saddle River, NJ 07458

Prentice-Hall International (UK) Limited, London
Prentice-Hall of Australia Pty. Limited, Sydney
Prentice-Hall Canada Inc., Toronto
Prentice-Hall Hispanoamericana, S.A., Mexico
Prentice-Hall of India Private Limited, New Delhi
Prentice-Hall of Japan, Inc., Tokyo
Simon & Schuster Asia Pte. Ltd., Singapore
Editora Prentice-Hall do Brasil, Ltda., Rio de Janeiro

Table of Contents

vi

Drawing and Photo List

Foreword

During over 50 years working in the electrical industry, the author commissioned many industrial electrical control systems in the USA and overseas. This required training the local electricians so they could understand and service the complex installations. The most frequent question was, "Where can we find a book that teaches us how to understand electrical logic diagrams?"

This book will give you the answers. It provides the generic training needed to help you analyze logic systems. It is a practical guide for electrical engineers, electricians, and full-time maintenance workers who are faced with real-life, on-the-job electrical control applications.

The material was first presented as a school to train professional elevator maintenance electricians working for VTS Inc, Seattle, WA. Each elevator manufacturer has his own unique control system, but all elevators still require equivalent control logic. This book deliberately uses non-elevator examples to provide objectivity. Older elevators use relay ladder logic, while the newer ones now use programmable controllers.

The books opens with a chapter which helps *define* logic, using abstract examples. We then define common components used to build logical systems. Simple logic and wiring diagrams in Chapter 3 illustrate where we are headed.

Chapters 4 through 9 take us on an adventure where we design a "sequencing" logic system around a fairly simple conveyor problem, develop the schematic diagram, make a bill of materials, and design component wiring diagrams. In Chapter 9 we go on site, and commission the installation, which includes some trouble-shooting experience to find problems that occur during the start-up. This shows how schematics and wiring diagrams can best be used to quickly solve problems.

Chapter 10 illustrates a case-history covering a "positional" logic system, complete with typical schematic and wiring diagrams.

Chapter 11 discusses samples of several unique solutions to problems that require bailing circuits, sorting systems, and counting with relays. This includes encoding and decoding with relays, and a basic shift-register circuit.

Chapters 12 through 18, provide a generic introduction to programmable logic controllers (PLCs). The subject matter does not try to make one a programmer, but defines the general layout of PLCs, the numbering systems one may encounter, basic memory structure, system addressing, and the common instruction set. The equivalent ladder diagram is shown, followed with an example of a program that might be used for the problem we previously solved in Chapters 4 through 9.

Chapters 19 and 20 introduce the contrast between digital and analog control systems, including an over-view of a DC motor, analog control system.

Final chapters recap several interesting trouble-shooting experiences and concluding remarks.

The appendices include a glossary, a summary of useful formulas and an index.

Our intent is to familiarize the electrician with relay ladder logic, and then show him how to understand the transition to PLC logic for similar installations. An electrician or technician who understands logic will find that he becomes a far more valuable trouble-shooter as he adds this expertise to his electrical experience.

Chapter 1

Introduction

Have you ever looked over a set of electrical drawings and wondered, "How can anyone make sense of all those confusing lines and symbols?" In reality, one will find that the most complex wiring diagram is an assembly of many simple circuits. Each circuit stores information or defines a condition that results in a specified action.

To understand the logic, one must be able to visualize what the machine or process is actually doing. One might ask, "What is a process?" One definition of a process is *a sequence of conditions that produces a desired result or product.* A process requires logic, and can be applied to ways of doing business, controlling an elevator, or the mechanical production of goods.

A mechanical process usually requires motors and/or hydraulic or air actuators, driving different parts of the process. There will also be conditional devices such as operator controls, limit switches, temperature sensors, proximity switches, counters, timers, etc., that control the sequence of events. A simple system may only require an operator's control station, where the sensor is the person's eye, and he logically starts and stops each part of the operation according to the needs of the process.

As a process speeds up, it is not practical to expect an operator to control it manually as fatigue and sheer boredom will take its toll. If there are multiple steps to perform, perhaps requiring timing decisions with another operator, the probability of errors will increase considerably. Errors result in loss of productivity as well as impacting the quality of the product.

World competition has forced industry to add automation to its processes, with the intent to eliminate as much human error as possible while improving productivity and quality. The result is that the operator has become a process supervisor. He no longer just watches the process, and pushes buttons to step the process through a sequence, but now supervises a complex, automatic system that he must be able to fine-tune and maintain for optimum performance.

Modern automation began with *relay logic*, where banks of interacting relays, controlled by system sensors, provided process *sequence* control. The circuitry is often designed as *ladder logic*. The schematic wiring diagrams have appearance of a ladder. Each line of logic appears as a rung of a ladder, between two side rails.

With the evolution of solid state components, many discrete mechanical hardware items were replaced by electronic devices. Some versions of ladder diagrams include *diode logic*, where DC control circuitry is directed through selected diodes. The diodes allow current flow in one direction and block any back-feed from conflicting circuits. The diode logic will often allow redundant use of some relays, thus saving space and hardware. See Chapter 3.

In the late 1970's, *programmable logic controllers* (PLC) appeared on the scene. This revolutionized the control industry, as one could now virtually eliminate banks of relays, as logic functions could be programmed in a microprocessor. The sensors are wired to the controller as inputs, and the motor controllers, actuators, etc., are wired to the output terminals. The printout of a typical PLC program follows the pattern of ladder logic, so the transition from discrete components to electronic controls is not difficult for anyone familiar with ladder logic.

Once you understand the logic of a system, you will recognize patterns of behavior that give you a feel for the process. Trouble shooting will become analogous to working a crossword puzzle where you may spot a missing link, even though you may not have known the word you were looking for.

Logic can follow many different paths so it is likely that your solution to a problem will be different from the next person. If we were to drive from one location to another in a large city, we might take different routes. If we arrive at the same destination, we are both correct in our logic to get there.

LOGIC

The term, *logic*, has been mentioned several times. What is logic? According to Webster's dictionary, logic is the science of reasoning. *Building upon known facts or conditions, we can predetermine the expected future event or condition.* The study of Plane Geometry is a good example of applying logic to solve theorems from known conditions. The same principle is the basis for building the sequential logic for electrical control systems. For each required action, we must determine the necessary conditions that must be true before the action can take place.

BOOLEAN EQUATIONS

George Boole, (1815-1864) an English logician and mathematician, developed a hypothesis that algebraic symbols could not only be applied to numerical calculations, but also to objects. The same algebraic calculations could be applied to add, subtract, multiply and divide objects to provide logical solutions.

Example: Let x = *horned* and y = *sheep*. Successive calculations using x and y would represent *horned sheep*. $1 - x$ would represent the operation of selecting all things in the world except horned things. $(1 - x)(1 - y)$ would give *all things neither horned nor sheep.* His line of reasoning expanded into trying to use this type of calculation to discover probabilities from known conditions.

Boolean equations are often used in university settings to represent complex electronic logic, but seldom used by "laymen". The electrical ladder logic format represents a form of logic that is very similar to Boolean logic, in that as you read it you will make statements such as *and, or, and not, or not* which relate to the functional relationship of each element, ending with a conclusion. No attempt will be made in this book to expand upon Boolean equations. We are just noting its existence so you may recognize it as another way of expressing logic.

EXAMPLE OF LOGIC

The first thing we must define is our goal. What is the required end result? Moving away from electrical applications, we will look at a hypothetical real-life situation. Pretend that you are barely fifteen years old and by the time you turn sixteen, you want to own a car. It appears that this is a common objective for teenagers in this day and age.

What steps must be taken before this goal can be realized? What are the obstacles that must be overcome? In other words, we must define the process and logic required to reach the expected outcome. Before continuing with this discussion, consider setting the text aside for a few minutes, and visualizing the dilemma that this fifteen year old is in. Write your own list of conditions.

Following are a few conditions that may have to be met.

    I   Parental permission
        A. Prove reliability
        B. Good school grades

    II   Drivers License
        A. Complete drivers education course
        B. Cost?
        C. Attain legal age to drive

    III   What kind of car do I want?
        A. Cost?
        B. Operating expenses
        C. Parking space requirements

    IV   Insurance requirements
        A.   Cost?

    V   Where do I get the money?
        A. Rich parents?
        B. Get a job?
            1.   What job?
                a.   Am I prepared for the job?
            2.   Will a job interfere with my study time?

Your list of requirements will probably be quite similar to the above if the recollection of your teen years is like the author's.

FISH-BONE DIAGRAM

A business consultant, Paul Webster, often uses what he calls the "Fish-bone" diagram, when he is advising struggling businesses. The head of the fish represents the goal, and he fleshes out the bones with steps required to reach that goal. We will use his analogy to illustrate this example. See Figure 1-1.

Starting at the tail of our "fish", we see that good grades and proof of reliability are the prerequisites to parental consent. The cost of the car, operating expenses, insurance, storage, driver's education and licenses are all factors to determine the budget. The license could be anywhere in the sequence, but requires driver's education and attaining the legal driving age.

Assuming that the car will not be an outright gift, the next step is to find a job, and complete any prerequisite training to qualify for it. The attainment of the job and collections of wages over the available time will produce money, which may be supplemented by a gift from the parents. Note that time may be a constraint as you still have to keep up the school grades. Sequence of events may vary depending upon circumstances.

This is a form of logic diagram. There are prerequisites for each step of the process as we work our way along the "back-bone". If grades fail, the process says that consent will be lost, and we start over. If all items are not covered in the budget, when we think the money is there, we will have to fall back and cover those costs.

THE LADDER DIAGRAM

Our car purchase logic can also be depicted as a ladder diagram. We have listed a number of obstacles that must be overcome. There are conditions that must be "true" before each obstacle can be overcome. The conditions can be depicted as normally open electrical contacts that close a circuit when proven true. When all required conditions are "true", the obstacle, represented by an output coil, is energized, and all future references to it become true. This pattern of logic will become clearer as you read Chapter 3.

Figure 1-2 illustrates the ladder diagram. The first rung on the ladder tells us that if we are reliable, and get good grades, we will have

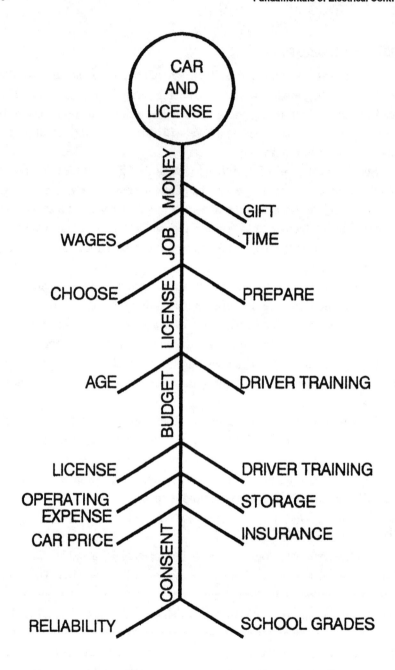

Figure 1-1 Fish-bone diagram

Car Purchase Logic

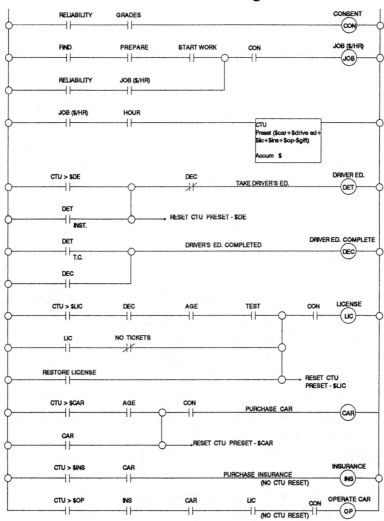

Figure 1-2 Ladder diagram

our parents' consent. Note that these requirements must be maintained at all times to keep "consent" energized.

The next rung tells us that we have to find a job, prepare for it, and work (with the consent of our parents). If we prove reliable, we will keep the job which represents dollars per hour of labor. This is represented by a sealing branch that keeps the job energized. Note that if consent is lost, due to failures in the first rung, the job can be terminated.

The third rung uses an up-counter *CTU* to represent a way of accumulating funds to meet our needs. If we have the job ($/hr), then we can increment our accumulation of dollars each time we work one hour. The preset value in the counter is the sum of all items requiring cash, less any gift that may be promised. When there is enough money to meet one requirement, it can be subtracted from the accumulation. If this is a one-time expense that is paid off, the counter's preset value can be reset to a value reduced by the need. The target is to accumulate dollars at least equal to the preset (need) value.

The fourth rung says that when our accumulated funds are greater than the cost of driver's education, we sign up for the course, and reduce the preset value of our counter. The course is represented by an on-delay timer. When you sign up, the course is instantly locked in until enough time has passed to complete the course.

When the course time is complete, the fifth rung is energized, indicating completion, and reaches back and cancels the class enrollment, by opening *DEC* (not complete) contact in the fourth rung. The driver's education is a requirement for the next obstacle, so it remains sealed-in as a permanent record.

Getting and keeping a license is a bit more complicated. The main branch of this rung requires that we have the cost covered, completion of driver's education, reached minimum age, passed driver's test, and still have parents' consent. If all these conditions are met, the license is locked in on the second rung. If we get a ticket, we can lose the license. It takes a restorative action to reactivate the license, provided that our parents still give their consent. Note that the conditions in the first rung still control the events at this point. We can also reset our counter now to a value reduced by the cost of the license.

Our next target is purchasing the car. Of course, the accumulated funds have to be equal to, or exceed the price of the car. There may be an age requirement, and of course, our parent's still have veto power. When you have bought the car, it is sealed in as an accomplished fact, and we reduce our target budget by the money used for the purchase.

We must have insurance before we can move the car. So now that we have the car, we must ascertain that there is money to buy insurance. When these needs are met, we are insured. Note that this is an ongoing expense, so we leave it in our budget in the *CTU* preset value.

Our *CTU* balance requires enough money for operating expenses. If we have insurance, the car, a license, and parents' consent we can now operate our car. Again, this is an on-going expense, so we have to maintain money in our CTU. In other words, we still need our job.

If our grades fail, we are grounded. If we lose our job, we park the car when funds are gone. If we get a ticket, we may be grounded by police or our parents. The logic requires that important conditions have to be maintained, not only to get the car, but to operate it after the earlier requirements have been met. Most will agree that we had similar conditions to face when we attained our first car.

THE FLOW-CHART

As another example, consider the simple logic of going through a door. If we were to teach a robot to go through a door, what decisions would it have to make? Following are some logical steps that might have to be completed.

I   Where is the robot?

II   Where is the door?
- A. Direction?
- B. Distance?

III   Is the door open or shut?
- A. Which way does it swing?
  1. Left?
  2. Right?
  3. In?
  4. Out?

IV Is the door latched or unlatched?
   A. Where is the latch handle?
   B. What motion must be made to unlatch?

V   What motions must be made to open the door and pass
    through?

This example is shown as a "flow-chart" diagram, illustrating the alternate choices to be made, depending upon the door configuration. See Figure 1-3 for a simplified version of the flow-chart logic. Obviously, the programming of the specific motions will be quite complex, but we can see the basic decisions that must be made to pass through the door. This logic should allow the robot to go through the door in either direction.

You can see by these illustrations that everything we do requires some logic. At some time in our life, we learned the logic of passing through a door. We are now "programmed" to make automatic observations, and respond to them with the required logical moves. If we make a faulty observation, we may bang our head if the door opens the wrong way – and we have to start over! As we begin to work with electrical logic, you will see that these concepts are valid whether you are designing logic, or trouble-shooting it.

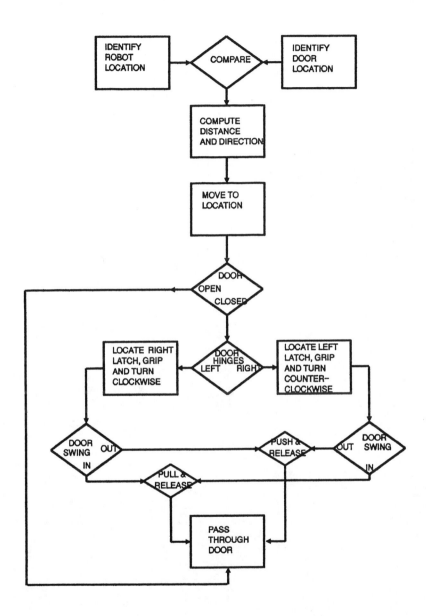

Figure 1-3 Flow chart

Chapter 2

Components

One of the first things we must do is identify the components that will be involved with our process. The components must be identified by their function in our documentation so we can understand what is actually happening as each component changes state. Following are the most common discrete components that will be encountered. Photographs have been provided, courtesy of the Allen-Bradley Co., Milwaukee, Wisconsin. The discussion includes some structural information to assist in proper application, and trouble-shooting.

CONTACTORS

Motors and other power consuming devices are usually controlled by contactors having heavy-duty contacts that can switch the power circuits safely. The contactor will be actuated by an electromagnetic solenoid (coil) which pulls the contacts closed when energized. They will also have an arc-suppressing cover to quench the arc formed when the contacts open under load. Ref: Figure 2-1.

If you are designing a system, be aware that AC contactors should never be used to break the current of a *loaded* DC power circuit. It is more difficult to quench the arc from a DC power load because there is no AC "zero crossing" to interrupt the current flow. DC contactors are designed to specifically handle DC current. You will find that they have imbedded magnets, or special blow-out coils, that are used to stretch the arc long enough to break the DC current flow.

Figure 2-1 Motor contactors with over-load protection.
(Courtesy Allen-Bradley Co., Milwaukee, WI.)

Devices using AC solenoids will have one or more single turn coil(s), called a *shading ring*, embedded in the face of their magnetic armature assembly. When the solenoid is energized, the magnetic field has continuous reversals matching the alternating current that is being applied. This will produce a noticeable hum or chatter because the magnetic structure is not energized continuously. As the magnetic field reverses, it induces a current in the shading ring, which in turn produces a secondary reversing magnetic field. The secondary field is out-of-phase with that of the applied power, and holds the armature faces sealed continuously between power reversals, and minimizes the noise.

Over time, shading rings tend to crack from the pounding of the armature faces. When this happens, the solenoid will become very noisy, coil current will increase, and premature failure will result. In an emergency, one can remove the damaged ring and replace it temporarily with a shorted turn of copper wire.

Figure 2-2 4-pole relay that can be modified for up to 12 contacts, or a timer attachment. (Courtesy Allen-Bradley Co., Milwaukee, WI)

RELAYS

See Figure 2-2. Relays have a similar construction to contactors, but since they are usually switching low-current logic signals, they do not have a requirement for the heavy-duty contacts and arc-suppression hardware. Most relay contacts have an AC continuous rating of no more than 10 amperes. They can close on an in-rush current of 150%, but only break 15% at 120 volts AC (vac). A NEMA A600 rating limits the inrush to 7200 voltamperes (va), and a circuit breaking rating of 720 voltamperes. As higher voltages are used, the current capacity goes down proportionately.

This difference in **make** and **break** ratings closely matches the typical ratio of inrush and holding currents of AC control coils. AC coils have relatively low *resistance,* allowing high in-rush currents. As the coil is energized, the AC current builds up *inductive reactance.* Total *impedance* (ohms), the vector sum of resistance and reactance, limits the continuous holding current. The ratio of in-rush to holding current is often 5 : 1 or more. Maximum impedance is attained when the air gap in the magnetic armature assembly has sealed closed. A failure to close this gap will reduce the inductive reactance, allow more current to flow, overheat the coil, and cause premature coil failure. A *shading ring* fracture will also lead to overheating and coil failure.

The same 10 amp contacts are only rated at 0.4 amperes DC, at 125 volts, and 0.2 amperes DC at 250 volts (50 va) because the small airgaps are not adequate to break a sustained DC arc. Voltages for DC logic controls seldom exceed 24 volts with typical current in the milliamp ranges.

Relays may have multiple coils for latching and unlatching of the contacts. Contacts may be normally open (NO), and/or normally closed (NC). When we speak of **normal,** it refers to the condition of the contacts when the relay is de-energized. The number of contacts usually varies from one to eight. Some relays use contact cartridges which can be converted for either NO or NC operation. When setting up NC contacts, one should try to have an equal number on each side of the relay so that spring-loading does not produce side thrust on the solenoid assembly. Most standard relays will have totally isolated contacts. Some miniature relays have type "C" contacts where a NO and NC contact share a common terminal. This construction requires careful planning to match the schematic wiring diagram to actual relay construction.

Occasionally the required load on relay contacts may be slightly higher than their nominal rating. One can increase the current capacity by connecting contacts in parallel, and improve the arc suppression by connecting multiple contacts in series. When multiple contacts are used in this manner, they should be on the **same** relay, because relay coils operating in parallel may have a slight difference in their time-constant. If one relay closes late, its contacts will take all the in-rush punishment. If a relay opens early, it will suffer more damage from breaking most of the full load current.

TIMERS

Timers are a type of relay that have pneumatic or electronic, timed contacts to meet various sequencing requirements. They may be "stand-alone" relays, or attachments to standard relays. *On-Delay* timers actuate the contacts at a preset time after the coil is energized, and reset instantly with power off. *Off-Delay* timers actuate the contacts instantly when energized, and reset at a preset time after de-energizing. NO and/or NC contacts may be time-closed, or time-opened. Many timers also include instantaneous contacts, actuated with no time delay. Instantaneous contacts may have to be added as a modification kit in some cases to stand-alone timers.

The coils of contactors and relays may be rated at a different voltage than the circuits being switched. This provides isolation between the low voltage logic circuits and higher voltage power circuits.

Figure 2-3 is a photograph of a pneumatic timing attachment that can be added to a standard relay. This timer can be arranged for either on-delay or off-delay functions, with NO and NC contacts.

Figure 2-3 Pneumatic timer attachment for standard relay. Convertible for time-on or time-off operation with NO and NC contacts.
(Courtesy Allen-Bradley Co., Milwaukee, WI)Milwaukee, WI)

Figure 2-4 Electronic timers with sealed contacts.
(Courtesy Allen-Bradley Co., Milwaukee, WI)

Figure 2-4 shows stand-alone electronic timer relays that use sealed contact cartridges that provide contamination protection and maximum reliability when switching low potential signals. These often have gold-plated contacts that can conduct and switch solid-state logic signals.

SOLENOIDS

Solenoids are used to actuate brake and/or clutch mechanisms, hydraulic valves, air valves, steam valves or other devices that require a push-pull motion. Some solenoids can become quite large, requiring contactors rated for their high current draw. Smaller pilot valves may draw no more current than a simple relay. It is important to pick the proper device to switch a solenoid.

Some heavy-duty units operate on DC current. The DC solenoids are often specified for operation in dirty or corrosive areas because current is controlled by circuit *resistance,* and will not rise if the air-gap is fouled. AC solenoids depend upon the *impedance* of the circuit. If the air-gap is not seated properly, inductive reactance is reduced and coil will draw excess current and over-heat as discussed in the relay section, above. Shading rings must also be watched for possible failures.

LIMIT SWITCHES

Mechanical limit switches have many configurations. Most will have both NO and NC contacts available. The contacts are switched when the lever arm is rotated a few degrees by a moving cam or slider. See Figures 2-5 and 2-6. The conventional drawing will show the con-

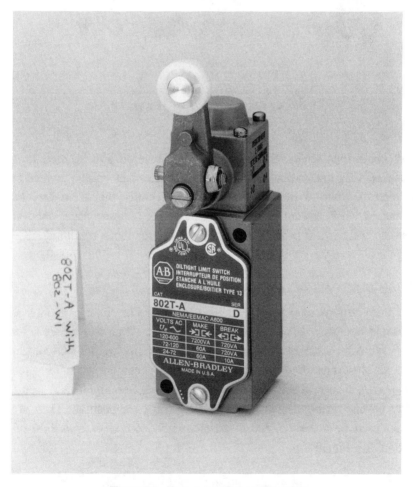

Figure 2-5 Typical mechanical limit switch.
(Courtesy Allen-Bradley Co., Milwaukee, WI)

Figure 2-6 Special purpose limit switches.
(Courtesy Allen-Bradley Co., Milwaukee, WI)

tact conditions when the machine is un-powered and at rest. It is assumed that the cam will normally strike the arm on the switch to change the state of the contact(s). You will notice that there may be instances where a cam holds its contact in its opposite state when at rest, so we have to provide separate symbols for those conditions. See Figure 2-7.

PUSH-BUTTONS

Push-buttons may have a single contact block, or an assembly of multiple contacts, depending upon the complexity of the requirement. The symbol shown in Figure 2-7 illustrates a single, NO/NC contact block. Most push-buttons have a momentary change of state when depressed, then return to normal when released. See Figure 2-8. Some may have a push-pull actuator that latches in each position. They are often used as a control-power master-switch with a *Pull-on/Push-off* action. See Figure 2-9.

Common Electrical Symbols

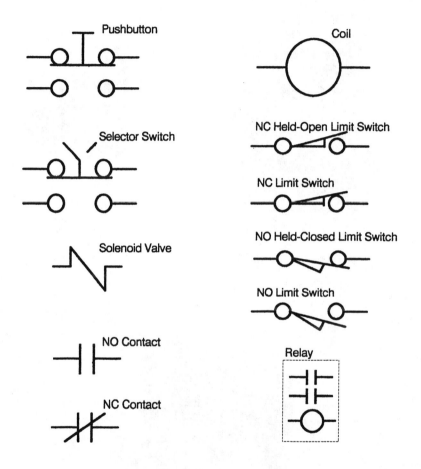

Figure 2-7 Common electrical symbols

Figure 2-8   Push-buttons (Courtesy Allen-Bradley Co., Milwaukee, WI)

Figure 2-9   Push-pull control button (Courtesy Allen-Bradley Co., Milwaukee, WI)

Figure 2-10 Assortment of 3-position selector switches
(Courtesy Allen-Bradley Co., Milwaukee, WI)

SELECTOR SWITCHES

Many selector switches have the same construction as push-buttons, except that the contacts are actuated by rotating a handle or key-switch. See Figure 2-10. The rotating cam may be arranged with incremental indices so that multiple positions (and contact patterns) can be used to select exclusive operations. The illustrated switch in Figure 2-7 has two positions, and the solid bar indicates the normal unactivated state. The cam mechanism can index at each position, or may be spring-loaded to return to a "normal position."

Contacts of push-buttons, selector switches, and limit switches usually have the same rating as the logic relays – 10 amps continuous at 120 vac.

NON-CONTACT LIMIT SWITCHES

There are a number of electronic "limit switches" that are used where it is not practicable to have an actuator arm physically contact a product or machine part. These switches include such items as *photocells,* and *proximity switches, (inductive, magnetic, capacitive).* In each case, a control signal is activated whenever an object enters its operat-

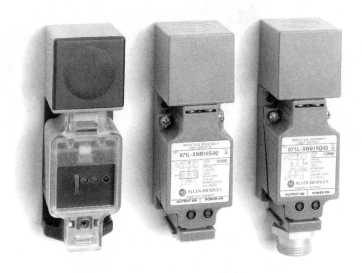

Figure 2-11 Non-contact, inductive proximity switches
(Courtesy Allen-Bradley Co., WI)

ing field. See Figure 2-11. These devices require additional wiring to energize their power supplies.

Relays or programmable controllers are often required to control power circuits that exceed the current rating of the device. Interposing relays or contactors are used to switch the higher powered circuits, thus amplifying the control capabilities of the low power device. They can all be shown in schematic wiring diagrams as labeled switches, contacts, and coils of relays and contactors. See Figure 2-7. If you are using a special device that doesn't lend itself to using conventional symbols, create your own symbol and provide proper identification.

Chapter 3

Ladder Diagrams

Ladder diagrams are the most common method of showing electrical logic. They are usually referred to as *schematic diagrams*, because they show the sequential scheme or order of events for the process and the control components required. The physical locations of the control components are not primary considerations in this diagram, only the sequence of operation and the relative timing (scheduling) of events. (The order of elements may be altered to simplify wiring requirements in some cases.)

Each rung of the ladder represents a separate event. The left sidebar is the beginning of the event, and the right one represents the completion. Electrically, the left side is usually the positive, or high side of the control power, and the right is the negative, or low side. The power source will be labeled at the top of the side rails, or shown connected to a source such as a control transformer.

Logic is read from left to right on each rung of the ladder. Conditions are represented by normally open (NO) or normally closed (NC) contacts. The resulting action, or stored information, will usually be represented by a relay, contactor coil, or solenoid. The conditional contacts are often referred to as *inputs,* and the coils as *outputs.*

Inputs such as push-buttons and limit switches are usually shown using their symbolic contacts.

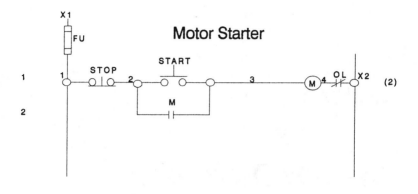

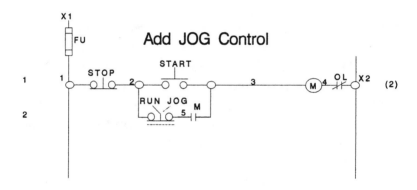

Figure 3-1 Motor starter schematics

The upper drawing in Figure 3.1 illustrates a logic rung for a simple motor starter. The normal "off" state is shown for each component of the rung. The *Stop* push-button (PB) is shown as NC, the *Start* push-button (PB) is NO, contact *M* is NO, the motor starter is shown as contactor coil *M*, and there is a conditional overload relay contact *OL*, which is NC. Since contact *M* has the same label as the contactor, it means that it will be actuated if the contactor is energized.

What does the logic of this rung say? "If *Stop* PB is **not** actuated, **and** *Start* PB **or** *M* **are** actuated, contactor coil *M* will be energized, provided that *OL* has not been tripped open. When coil *M* is energized, contact *M* will close to maintain the circuit when *Start* PB returns to its normal state. If either *Stop* PB or *OL* are actuated, *M* will be de-energized." Note that contacts in series are "and" statements, and those in parallel are "or" statements when changing state from normal to actuated. When returning to a de-energized state, any series contact will open the circuit, so an **or** statement would be true, and any parallel contacts would require all to be open, so an **and** statement would be correct. Verbalizing a rung's logic is sometimes called a *Boolean statement.* Boolean algebra, as discussed in Chapter 1, is another way to depict logic, and those familiar with it will recognize the relationship to ladder logic.

You will also notice that conventional labeling is shown in Figure 3-1. The side-bars are labeled X1, X2 indicating control transformer terminals. Each rung level is numbered on the left. Each segment of the rung is numbered – all elements that a segment touches will have a common terminal number. On the right, the marginal number indicates that contactor M has a NO contact on line 2. If there were a NC contact, it would be underlined.

This is the most widely accepted format at this time because it lends itself to reasonably easy interpretation. It is shame that many vendors do not bother to provide these simple documentation helps so maintenance people can quickly understand their logic.

Let's add another bit of logic to Figure 3-1. There are times when we may wish to have the motor run only while we hold down the *Start* button, stopping when it is released. This is called **jogging**, and is often used to move a conveyor to a starting position, provide control to clear an obstruction, or test motor operations.

The reason that the original circuit "seals-in" is that contact *M* bypasses the *Start* button as soon as contactor *M* is energized. If we can keep this part of the circuit open, the motor will only run while *Start* is activated. One method of accomplishing this is to add a selector switch *(Run-Jog)* in series with the *M* contact. The motor will then run when the selector switch is closed, and jog when it is open. See the lower drawing in Figure 3-1.

WIRING DIAGRAM

After developing a *schematic diagram*, the next step is to build a *wiring diagram* that shows the electrician how to physically inter-connect the components. All labeling and wire numbers developed in the schematic diagram must be carried over to this diagram. If the device has permanent terminal numbers marked on it that may not match the schematic, they may be noted, but the wire number must follow the schematic. We will use the motor starter schematic, Figure 3-1, for an example. A complete schematic drawing would also include the power circuits. In this case we did not show them, but will include them in our wiring diagram example.

Assume that we are using a combination starter than includes its own circuit breaker and control transformer. The push-buttons are at a remote operator's station. See Figure 3-2. Power wiring enters the motor starter enclosure from a 460 volt, 3 phase source, connecting to terminals *L1, L2,* and *L3* on the circuit breaker. A control transformer, *Ctx* is connected after the circuit breaker, and may be pre-wired from the factory. High voltage fuses may or may not be in the transformer input, depending upon code requirements. The logic gets its power from control transformer *Ctx,* terminals *X1* and *X2*. A terminal strip provides connections for the three control wires that run to the remote control station. This illustrates the circuit which is usually used for "three-wire control."

In Figure 3-3, you will see that we are able to add the *Jog* selector switch to the wiring diagram without adding any extra wires between the contactor and the operator control station. The device was added to the remote station and the wire previously labeled #2 returns to the starter as #5. If factory terminal markings do not match the schematic, it is important to re-label as necessary.

Notice that in the schematic drawing (Figure 3-1), that *Run-Jog* is shown ahead of the *M* contact. If the contact had been shown ahead of the selector switch, we would have needed two additional wires between the starter and the control station. [As an exercise, draw the schematic with these two components in reverse order, and then develop a wiring diagram to show minimum wiring required].

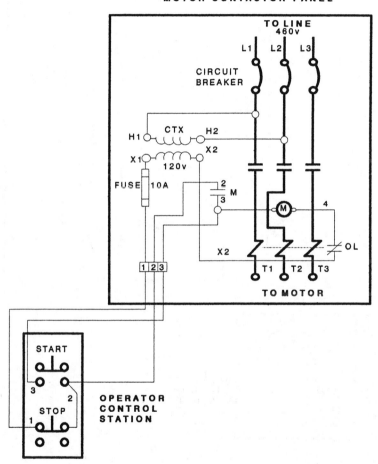

Figure 3-2 Motor starter wiring diagram

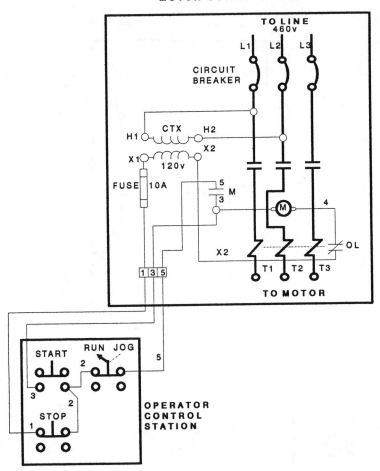

Figure 3-3 Motor starter wiring diagram with jog function

This illustrates the importance of organizing your logic to keep the necessary inter-wiring as simple as possible, saving material and labor costs. In some cases, other circuits might be included in the same conduit run, and the extra wiring might require a larger conduit size.

The wiring diagram closely matches the physical wiring of each component, and shows the necessary wiring runs to remote devices. It is, much harder to trace the logic from this diagram because of the physical routing of the conductors. Since the labeling matches the schematic drawing, it is easy to locate each component that was shown in the schematic.

Large installations require multiple wiring diagrams to show the components, so one might have to use several wiring diagrams to follow a single line of logic through all its terminal connections.

DIODE LOGIC

There are occasions where ladder logic can be simplified by including diode logic to reduce the number of components in a circuit. The control power in this case will usually be from a low voltage DC source. Diodes are rectifiers that allow the current to flow in only one direction, and since they block reverse current, they can be used for signal "steering".

The upper drawing in Figure 3-4 shows the relay logic that might be used to select alternately one, two, or three relays from a single three-position selector switch. This circuit might be used to set up three stages of heating, lighting, or a group of motors. Note that auxiliary relays are used to provide the extra contacts to prevent backfeed between selected relays. A total of six relays and six contacts are required in addition to the selector switch.

The lower drawing in Figure 3-4 shows the same logic, using six diodes and three relays, directly controlled by the selector switch. A single relay can be controlled from multiple sources. This is a method of simplifying the circuit and still prevent back-feed between control circuits without using extra isolation contacts. Note that a DC source is indicated for the diode logic.

Visualize the selector switch in each of its three positions, and trace the three conditions in each logic diagram. You will find that in the

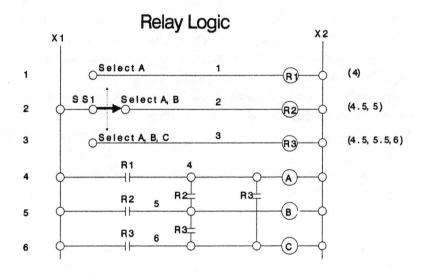

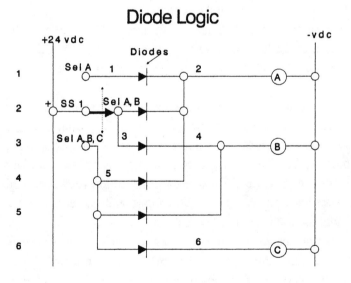

Figure 3-4 Compare relay and diode logic

upper position, only relay A will be energized, and B and C cannot be energized. In the middle position, as shown, relays A and B will be energized, and C cannot be energized. In the lower position, all three relays, A, B, and C, will be energized. Both logic diagrams accomplish exactly the same end result, so both can be considered correct. You may come up with a different circuit that would also be correct, if it accomplishes the same results.

When one understands the logic of a circuit, it is easiest to trouble-shoot the system from the schematic diagram. Using the diode logic diagram, imagine that when we select a single relay or all three relays, everything works fine. If B fails to energize when selecting two relays, we know that the diode between wires 3 and 4 has an open circuit, or a bad connection.

Trouble-shooting should always start from the schematic diagram to avoid the confusion of trying to follow a wiring diagram through all its inter-connections. Once the fault has been identified from the logic, one can locate the discrete components in the wiring diagrams, and make necessary corrections.

This chapter has illustrated some basic concepts for conceiving a logical schematic diagram and a simple wiring diagram. It is obvious that most processes are a lot more complex than shown, but just remember that each rung is a single line of *logic*, with conditional inputs that produce an expected output. The inputs may be derived from remote inputs or the output of other lines of logic, so it is important to setup a system of labeling that provides good identification for each component. It also useful to annotate each rung to identify its purpose, such as "Start pump #1", "Open control valve #4", etc. This can be written above or below the line of logic in the ladder diagram.

Another tip is to keep the logic as simple as possible to keep the "confusion factor" at a minimum. Following is a case history illustrating the importance of keeping the logic simple.

CASE HISTORY

Several years ago, an Oregon plywood plant was having a lot of trouble with a control system that loaded uncured plywood into a multiple-opening hot press. The machine was designed to automatically push one platen, carrying the raw plywood, into each opening of the press, then index to the next level and repeat. Occasionally, the system would "forget" where it was and jam two loads into the same slot, creating a real mess.

The production manager complained bitterly to the press manufacturer because he was damaging machinery and product, and of course interrupting production. The manufacturer finally engaged the author to take a fresh look at the logic his engineer had developed. We changed the shape of one cam, added one limit switch, and removed twenty (20) relays from his circuit.

The logic of the original schematic was correct – if everything maintained the proper sequence. The problem was that they were experiencing "racing relays" where the circuit required simultaneous operations. Varying temperatures will change a relay's time constant. In this case, position-indicating relays occasionally dropped out during a period when their control was supposed to be instantly transferred between two interposing relays – whose time-constants were apparently changing.

The cam on the elevator drive originally had two detents that actuated the same limit switch alternately as it arrived at even and odd numbered levels. This pulse alternately switched relays between even and odd numbered levels each time a detent passed the limit switch. Since the circuit had to "remember" where the elevator was at all times, any relay that "mis-fired" caused it to lose its place.

By making a new cam with one detent that alternately actuated "even" and "odd" limit switches, we had a positive indication of the position at all times, and no longer needed the bank of "racing" position relays.

The problem with the original circuit was worsened by additional relays the engineer had added to "compensate" for the apparent fault. We are not inferring that the engineer was incompetent, he had just become too involved with his original logic to look at it objectively. This illustrates the point that all of us occasionally "cannot see the forest for the trees", and should consider asking someone else to take a fresh look at a problem that has us puzzled. There should be no shame in having someone look over your shoulder when you hit a snag in the system you are working on. In this case history, a simple change in the input hardware allowed us to <u>remove</u> the problem, instead of correcting it with more circuitry.

In the following chapters we will use the above principles to develop schematic and wiring diagrams from "scratch". Take your time as you work through those chapters as there are many references to drawings included in the text. The chapters will be relatively short, to encourage taking time to understand each step as it is presented. Remember that you may have solutions that differ from those shown in the text, and if your logic works, it is just as correct as the author's.

Chapter 4

Develop a Schematic Diagram

We will now develop a schematic diagram around a specific application. The first thing we have to do is ask questions! One cannot ask too many questions.

- What is the application supposed to accomplish?
- What are the principal power components?
  - Motors?
  - Brakes?
  - Solenoids?
- What is the power source?
  - 3-phase, or Single phase?
  - Available voltages?
- What are the required modes of operation?
  - Manual?
  - Automatic?
- What protective circuitry must we provide to prevent hazardous operation?
  - Safe start-up?
  - Safe stops?
  - Warnings, alarms?

- What control voltage should we use?
  — 120vac, 24vdc, etc.?
- What environmental conditions will we have to contend with?
  — Dusty?
  — Wet?
  — Hazardous dust, gases, corrosives?

This is a starting point. The operators may have specific requests for multiple control points, choice of equipment, color-coding of equipment, etc. All these requirements should be determined as early as possible to prevent costly re-design and re-work.

The following example will help us work through the process of developing a schematic ladder diagram. Some of the questions, above, may not apply but never-the-less would be part of our preliminary evaluation of any project.

DESCRIPTION OF SEQUENTIAL APPLICATION

We have a grain elevator that loads grain from its silo(s) into a ship. Because of the site configuration, and ship positioning, a series of three (3) conveyors are required. See Figure 4-1. Our responsibility is to provide a safe operating sequence for the conveyors, C1, C2, and C3. In real life, we would probably incorporate silo gates and diversion controls in the total system, but in this exercise we will limit our schematic to the conveyor controls.

Power will be supplied at 460v, 3-phase, 60 Hz at a central motor control center. Each conveyor will have a single speed, non-reversing 50 Hp motor, supplied by others. Controls will be at an elevated control cab, overlooking the conveyors. Conveyors are each about 100 feet long, operating at 400 feet per minute.

Three modes of operation are required:

Automatic mode

- Start in a timed sequence of C3, C2, C1 to prevent dropping grain on a stopped and/or loaded conveyor.
- Stop in a timed sequence of C1, C2, C3 to allow the system to totally empty.

Conveyor System

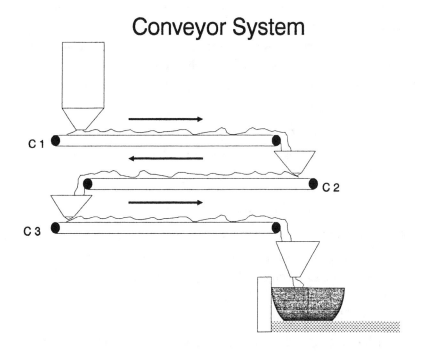

Figure 4-1   Conveyor system plan

Manual mode

- C3 must start first.
- C2 cannot start if C3 is stopped
- C1 cannot start if C2 is stopped
- If any conveyor is stopped, the infeeding conveyor(s) ahead of it must stop to prevent piling grain on a stopped conveyor.

Manual Jog mode

Manual start buttons can be switched to jog duty using a selector switch. In this mode, a conveyor will only run while the start button is held down, and stop when released. Jog can be used on any conveyor regardless of starting sequence. All conveyors must stop running and be in a jog mode when this is selected. This allows for intermittent operation for emergency clean-up, belt alignment, etc..

Since there is the possibility of having maintenance people working around the conveyors, out-of-sight of the control cab, it is necessary to have a 15 second audible warning before the start-up of the conveyors in the automatic mode. We will provide relay contacts to control a warning system, supplied by others. The following control sequence is required:

- Select *Automatic*
- Momentarily press *Auto Start*
  — Audible warning turns on immediately and continues
  — At 15 seconds, C3 starts
  — At 30 seconds, C2 starts
  — At 45 seconds, C1 starts and warning stops.

For an automatic stop, the following sequence is required:

- Momentarily press Auto Stop (Assume silo gate is closed)
  — After 15 seconds, C1 stops (has unloaded)
  — After 30 seconds, C2 stops
  — After 45 seconds, C3 stops

Each conveyor will illuminate a respective *green* indicating lamp at the console when running.

A *red* indicating lamp will turn on whenever any motor overload trips, and all conveyors will stop instantaneously.

A *red* emergency stop push-button (E-Stop), in the console, will stop all conveyors instantaneously. (Additional E-stop buttons can be added in series with this push-button to provide additional remote emergency control.)

Switching between *Manual* and *Automatic* modes must be made with the conveyors stopped. Selector switch must have a center OFF position.

POWER LOGIC

Since we are going to control three (3) 50 Hp, 460v, 3 phase motors, we will have to provide necessary disconnects and starting contactors to meet National Electrical Code (NEC) standards. We need to show that our power source is 460v, 3 phase. The disconnects can be fusible switches, or circuit breakers (CBs).

For our example, we will choose CBs for our disconnecting means. Each contactor, in the motor control center, includes its CB disconnect, power contacts, a 3 phase overload relay, and terminations for its respective motor. The contactor and CB ratings will be determined during the workup of our final bill of materials. Labels C1, C2, and C3 are chosen to indicate the respective conveyor motors, and will be carried forward for contactor labeling.

Our control power must also be derived from this 460v source. For safety reasons, the control power will be supplied at 115v, single phase. This will require a 460/115 volt control transformer with necessary disconnecting means and secondary protection. The Kva rating of this transformer is unknown at this time because we do not know how many relays, indicating lamps, etc., that our schematic will require. See Figure 4-2.

The transformer terminals are labeled with their conventional markings, H1-H2 for high voltage terminals, and X1-X2 for the low voltage terminals. As a rule, X1 will be the high side of the control circuit, and X2 will be the grounded, low side of the system.

On the left margin of our schematic, we begin line index numbering, 1, 2, 3, etc., to help locate components that may be referred to in the schematic or instruction manuals. (Under-lined numbers are used to identify line indices for text only.)

- Lines 1, 2, 3, show the AC power source with phases A, B, and C. labeled "460vac, 3 phase".

- Line 4 shows all 460 volt disconnects.

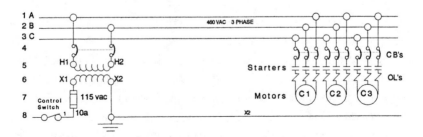

Figure 4-2 Power logic

- Line **5** shows the control transformer 460v winding and terminations, and the motor starter contacts.

- Line **6** shows the control transformer 115v winding and terminations, and the motor overload relay elements.

- Line **7** shows control transformer secondary fuse, and the conveyor motors.

- Line **8** shows the control circuit switch, and grounded X2 terminal.

Although we have compressed this portion of our schematic to conserve space, this part could justify a separate drawing, especially if we had to include other lighting and power loads. Note that schematically, each motor is connected directly to its overload relay terminals, but in reality those leads may be several hundred feet long.

SAFETY LOGIC

At this point, the power schematic is complete. The next problem is to develop the logic required to control the power. In the description of operation we start by looking for conditions that prevent operation, or shut down the entire system. These are safety related conditions that must be satisfied before any motor can be started.

Three conditions were listed:

1. A *red* indicating lamp will turn on whenever any motor overload trips, and all conveyors will stop instantaneously.

2. A *red* emergency stop push-button (E-Stop), in the console, will stop all conveyors simultaneously.

3. Selector Switch must have a center OFF position.

Each motor contactor has its own overload relay with a NC contact that will open when the relay trips from over-current (heat). It is usually prewired to the low side of the contactor coil. Since we want *any* overload to stop *all* motors, it is necessary to isolate the contacts from the original connections and combine their functions in a single relay circuit. The three NC contacts can be connected in series to hold in an overload relay when conditions are normal, and any one of them can be used to de-energize the relay, and indicate that a contactor overload contact has tripped. How can we use this in our schematic?

We will label this relay *OLR*. Our logic statement for this relay: "Operation will be permissible when *OLR* is energized, and no operation will be permissible if it is de-energized. A *red* light will be lit when it is de-energized, indicating a tripped overload."

Our previous schematic, Figure 4-2, ended with a control switch to feed our logic, and line X2 as the grounded, low-side of line. We now continue our circuit numbering by assigning wire number 2 to the left ladder rail. Now draw in the three NC contacts *(OLC1, OLC2, OLC3)* in series with *OLR* between lines 2 and X2. Our red indicating lamp circuit is made up of a NC contact *OLR* and the lamp R, connected as a parallel circuit between 2 and X2. A NO contact *OLR* is added to the continuation of the left ladder rail, and serves as a safety shut-off for all the circuitry to follow. If *OLR* is not kept energized by the overload contacts, the total system will shut-down.

Figure 4-3 shows this circuitry. We have shown the lamp circuit above the relay as a conventional arrangement, but placing it below would be just as correct. The wiring segments have been numbered in sequence, left to right and top to bottom. The left rail becomes wire number 7 below the NO contact *OLR*. Rung numbers 9 through 11 have been added to the left of the drawing, and contact indexing on the right

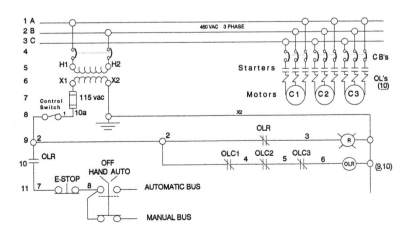

Figure 4-3 Power and Safety logic

shows where *OLR* contacts are used. The underlined contact index number indicates a NC contact. The power schematic shows an index for the overload contacts.

The second required safety device is the *E-Stop* push-button. By adding it in series with the NO contact *OLR*, we have met the requirement. If either *OLR* is de-energized or *E-Stop* is pressed, the control circuit will be interrupted, and nothing can be operated "down-stream" from this point. If additional *E-Stop* push-buttons are needed at strategic places near the conveyors, they can be added in series with this push-button without limit.

The third requirement is a center OFF position for the *Hand/Auto* Selector Switch. It is added in the circuit after the *E-Stop* push-button and the connection points for Automatic and Manual control circuits are identified. We will work out separate logic patterns for the *Hand/Off/ Auto* circuitry, and add it later to these points.

Chapter 5

Logic Continued

In the last chapter, we developed our logic that is related to all operating conditions. It is called *permissive* logic because all conditions must be satisfied in this part of the schematic before any Automatic, Manual, or Manual Jog control can operate, and it can over-ride the control at all times.

At this point, we do not know how many relays will be required, nor do we know how the functions will be interlocked. Instead of adding directly to the original schematic, we will use a "scratch-pad" to work up each function separately and fit the results into the drawing later.

AUTOMATIC START

Since the *Auto Start* and *Auto Stop* push-buttons are momentary contact devices, we need to "latch-in" the control circuitry so that the required operating cycle can be completed. You will note that the start cycle requires a warning signal that begins 15 seconds before conveyor C3 starts and continues until C1 starts. This means that we will have to unlatch the warning, and at the same time maintain C1, C2, and C3 to keep them running.

On the left side of our scratch pad, we will label a side-rail "automatic bus". Our first element will be an *Auto Start* push-button that must immediately energize our warning. The warning could be a bell, horn, flashing light, or even a siren, but for our purposes we will show a warning relay coil W whose contacts can be used for any device. The

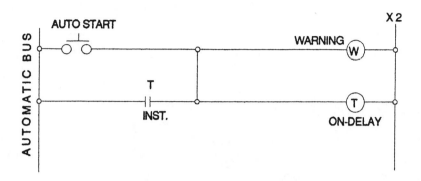

Figure 5-1 Auto-cycle warning circuit logic

circuit is extended from the push-button to W and then to the right side-rail labeled X2. W must be held energized during the entire start cycle, so it will require a sealing contact from some source. We also want to start our timing cycle at the same time. A simple solution is to wire an on-delay Timing Relay T in parallel with W, and use a NO instantaneous contact T to by-pass our push-button and seal in both the warning and the initial timer.

The alarm must operate 15 seconds before C3 starts, so T will have its time contact set to close 15 seconds after it is energized. Its instantaneous contact will close immediately and stay closed until T is de-energized. See Figure 5-1.

We might be tempted to have T start C3 directly, but remember that timing for C2 is dependent upon C3, and we also have to allow for a separate manual control of all conveyors. We can use T to start the next time cycle by energizing an Automatic Timer for conveyor 3, AT3. Note that we label the relays to identify their function if possible.

AT3 will need a 15 second, on-delay contact to start the C2 cycle, and an instantaneous contact to start and maintain C3. See Figure 5-2.

We can continue this logic by using time-delay contact AT3 to energize an on-delay timer AT2, having a 15 second, on-delay to start the C1 cycle and an instantaneous contact to start and maintain C2. The AT2 timed contact will energize A1 which has a normally open contact to control C1. Note that A1 does not need to be a timer.

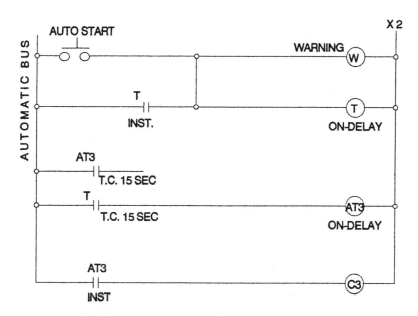

Figure 5-2 Auto-cycle, add first conveyor start logic

At this time we want to turn the warning off, so we can insert an *A1*, NC contact in the *W* circuit to turn off the warning when *C1* has started. See Figure 5-3.

The specifications state that each conveyor must show a green indication lamp when turned ON. If we connect green indicating lamps *G1*, *G2*, *G3* in parallel with the respective contactor coils *C1*, *C2*, and *C3*, the lamps should light as each conveyor starts in sequence. In reality, with this arrangement, the lamps only indicate that the *C1*, *C2*, *C3* contactors have been commanded "on." If we need to <u>prove</u> that the contactors are actually energized, and "pulled-in," we need to include a NO contact *C1*, *C2*, *C3* in series with each respective lamp. Remember, the motor control center may be located hundreds of feet away from the operator's console.

Now trace the logic in Figure 5-3. Momentarily close the *Auto Start* push-button. Relays *W* and *T* will immediately energize, seal in, and warning will start. 15 seconds later, *AT3* is energized, and starts *C3*. After another 15 seconds, *AT2* is energized, and starts *C2*. Then after another 15 second delay, *A1* is energized to start *C1*, and turn off the warning.

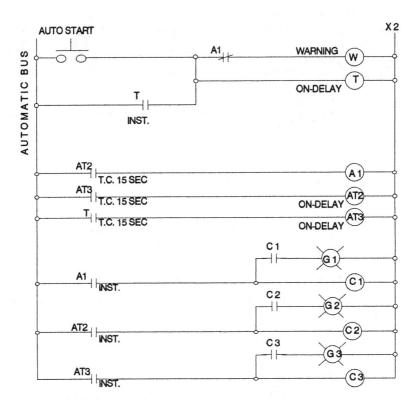

Figure 5-3   Auto-cycle, complete conveyor start logic

AUTOMATIC STOP

By installing a NC *Auto Stop* push-button in series with the instantaneous *T* contact, we can break the seal, and all conveyors would come to an immediate stop. However, we are required to stop in a reverse sequence to the start cycle, so that all conveyors will emptied during a normal automatic shutdown. We are assuming that the stop cycle will begin about the time the silo supply is turned off. This means that all conveyors must be held in a running mode for varying time periods after *Auto Stop* is initiated.

This stop delay can be accomplished by adding three, off-delay timers which will begin timing out when *Auto Stop* push-button is momentarily pressed. These Stop Timers *ST1, ST2, ST3* are connected to the start "seal" circuit, and energized during the start mode. A NO, 15 second off-delay *ST1* contact is inserted in the A1 circuit; A NO, 30 second off-delay *ST2* contact is inserted in the A2 circuit; and a NO, 45 second off-delay *ST3* contact is inserted in the A3 circuit.

Now we run into a problem. When *Auto Stop* was initiated, *T* was de-energized, which would let *AT3* to drop out. If *AT3* is de-energized, *AT2* and *A1* will "avalanche" off and bring all conveyors to an immediate stop. To correct the problem, and hold *AT3* energized, we will add a NO, instantaneous *AT3* contact in parallel with the original 15 second, time-close *T* contact. *AT3* is then no longer dependent upon *A1* to hold it energized. See Figure 5-4.

We did not change the Start cycle, but now when we initiate an automatic stop, *C1* will stop after 15 seconds, *C2* will stop after 30 seconds, and *C3* will stop after 45 seconds, allowing time for all conveyors to empty.

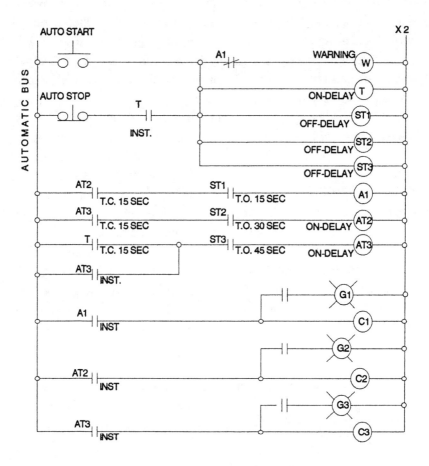

Figure 5-4 Auto-cycle, add auto-stop logic

Chapter 6

Logic Continued

MANUAL CONTROLS

Now we have to add the necessary logic to allow manual control of *C1, C2,* and *C3* as an alternative to the automatic sequences determined in Chapter 5. This time, we will label the left side of our schematic as the *Manual Bus.*

A simple way to develop the *manual* circuits is to draw typical *Start/Stop* controls for each motor using relays, instead of the motor contactors. The relay circuits can be set up to provide necessary control interlocking, then the result will be used to control the contactors. See Figure 6-1.

We have labeled the relays *M1, M2, M3* to control *C1, C2, C3* respectively. You can see that the circuits are the same as the upper illustration in Figure 3-1, except there are no *OL* contacts. That arrangement would provide logic for starting and stopping the motors, but does not provide the necessary interlocking that requires *C3* to run before *C2,* and *C2* to run before *C1.*

By inserting an M3 NO contact ahead of the M2 coil, and an M2 NO contact ahead of the M1 coil, we can force the required manual start sequence. See Figure 6-2.

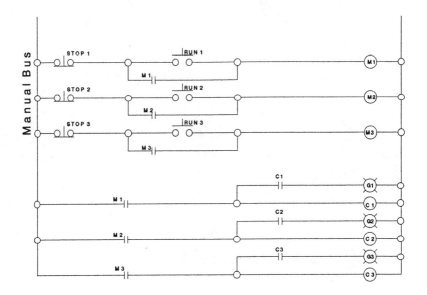

Figure 6-1 Manual-cycle conveyor run logic

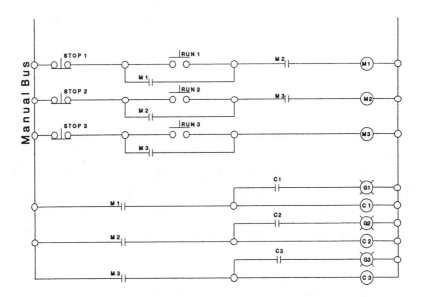

Figure 6-2 Manual-cycle, add stop sequence logic

Note that with this arrangement, any time a *Stop* PB is actuated, all conveyors above it will stop automatically, preventing an over-run of materials onto a stopped belt.

JOG CONTROLS

Manual operations also require *Jog* control. We can emulate the lower circuit in Figure 3-1 to add this feature. Since there are three relays to set up, we will install a jog relay JR, controlled by a *Run/Jog* selector switch, and use its contacts, as needed, to modify our control circuits.

By installing a *JR* NC contact in series with each of the sealing contacts *M1, M2, M3,* we prevent the relays from sealing in. Since *M2* has an *M1* NO contact ahead of its coil, and *M1* has an *M2* NO contact ahead of its coil, it is necessary to add *JR* NO contacts as by-passes to those contacts to get a signal to the respective coils. Note that *JR* requires a total of three (3) NC, and two (2) NO contacts. See Figure 6-3.

We can now test this circuitry. If we set our selector switch to *Run*, we find that *M3* must be started before *M2*, and *M2* must be started

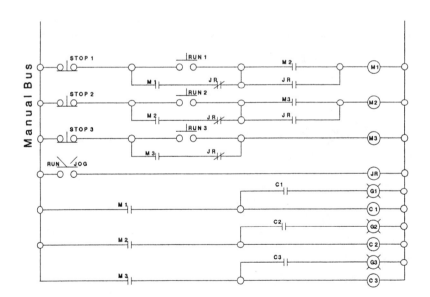

Figure 6-3 Manual-cycle, add jog logic

before *M1*. If *M2* is stopped, *M1* will also stop. If *M3* is stopped, both *M2* and *M1* will stop. If the selector switch is set to *Jog,* the respective *Start* PB can jog its relay without respect to any start sequence. Note that when *Jog* is selected all other motor control relays will be de-energized immediately if previously sealed in.

PUTTING IT ALL TOGETHER

We have developed three blocks of logic that have to be combined. Figure 4-3 is the beginning our schematic, and provides basic power and permissive circuits. Figure 5-4 is our scratch-pad schematic of the automatic logic, and Figure 6-3 shows our manual logic.

Figure 6-4 illustrates how these blocks of logic can be combined into a single schematic diagram. Note that wire numbers have been added, and contact indexing is shown on the right of the drawing to illustrate how many contacts each relay requires, and where used. The *C1, C2, C3* contactors are controlled from two independent relay sources. The *Hand/Off/Auto* selector switch prevents any possibility of both sources being energized at the same time.

Notice that we have divided the circuitry into five (5) functional logic blocks, making it easy to trouble-shoot the logic systematically.

1. Power sources
2. Permissive, safety logic
3. Automatic logic
4. Manual logic
5. Combined output control.

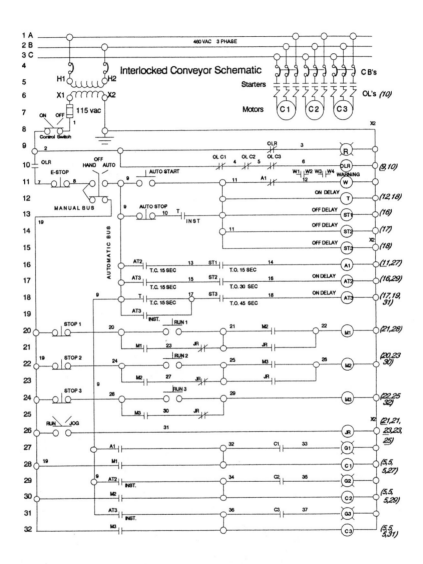

Figure 6-4 Completed conveyor schematic

Chapter 7

Develop a Bill
of Materials

Now that we have developed our schematic diagram, we need to identify the hardware that will be needed for our project. One item of particular interest is the volt-ampere (va) rating of the control transformer. This requires that a specific bill of materials must be developed, and then determine the worst case transformer load. This load will be composed of the *inrush load of the largest contactor*, plus the *holding* load of the additional contactor coils, relay coils and lamps that can be energized at the same time.

In many industrial installations it is a common practice to install motor control centers for multiple motors connected to the 460v, 3 phase power source. Modular plug-in power modules are available to provide combination starters, control transformer components, etc. as required. See Figure 7-1. The three (3) 50 Hp combination motor starters, *C1, C2, C3* may be installed in a control center with many other motor controls that are unrelated to our specific application. Catalog information for "open type" starters can be used to get the coil current ratings to help determine the load on the control transformer.

There are many brands of starters and relays available, but for our example we will use data from an Allen-Bradley industrial catalog. An open type 50 Hp starter, with 120 volt control, will use a NEMA size 3 contactor, catalog number 509-DOD. See Figure 7-2. Its coil will have

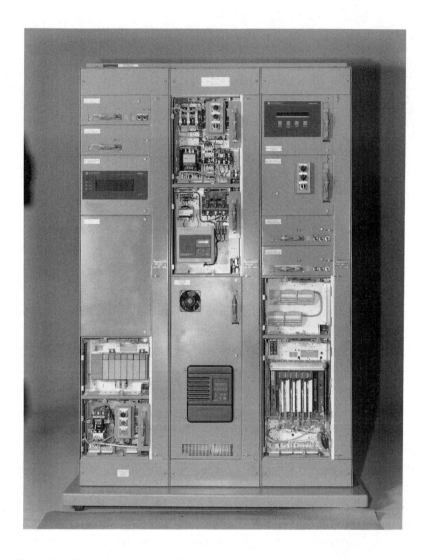

Figure 7-1   Motor control center, showing a wide variety of plug-in modules, including AC drive, PLC5, SLC 500 programmable control, motor starters, and various "smart" devices. (Courtesy Allen-Bradley Co., Milwaukee WI)

Figure 7-2 NEMA size 3, motor contactor (Courtesy Allen-Bradley Co., Milwaukee WI)

an inrush of 660 va, and a sealed rating of 45 va. Since this is the largest load, we will allow 45 va each for two contactors, plus the inrush of 660 va for one contactor as the maximum load from the power components. This totals 750 va.

There are also 13 relays and 5 indicating lamps to consider. We will now develop that list of materials.

Relay	NO contacts	NC contacts	Cat. No.	Hold VA
OLR	1	1	700P-110-A1	20
W	(allow) 2		700P-200-A1	20
T	1, 1 TC-15s		700PT-200-A1	20
ST1	1, 1 TO-15s		700PT-200-A1	20
ST2	1 TO-30s		700PT-200-A1	20
ST3	1 TO-45s		700PT-200-A1	20
A1	1	1	700P-110-A1	20
AT2	1, 1 TC-15s		700PT-200-A1	20
AT3	1, 1 TC-15s		700PT-200-A1	20
M1	2		700P-200-A1	20
M2	3		700P-300-A1	20
M3	3		700P-300-A1	20
JR	2		700P-230-A1	20
Red lamp			800T-P16R	1
Green lamps (3)			800T-P16G (total)	3

If we study the logic diagram, one will find that the maximum number of relays that can be energized simultaneously is eight (8), and there will never be more than three (3) lamps energized at any time. The worst case instantaneous contactor load was 750 va, and the relay and lamp load totals 163 va, so our total momentary maximum load is 913 va. If we consider that the continuous sealed load for contactors is only 135 va (45 x 3), the total sealed load, with relays, would be 298 va.

We now have to make a judgment call. Most control transformers can momentarily provide up to 5 times their continuous rating for momentary overloads. If the inrush is too high for our transformer, other contactors may drop out from the excess voltage drop. In this instance we can use a 500 va transformer because of the maximum momentary load of 913 va is less than 200% of our transformer rating. If we choose the 500 va transformer, the catalog number will be 1497-N19. This can be fed from 10 amp, or smaller, circuit breaker, from the 460 volt source. It may be prudent to expect other devices to be added to this circuit, possibly by others, so we might choose to increase the transformer size to 750 va to cover future contingencies. The cost is really not that much more.

In addition to the above items, we also have to pick out the push-buttons and selector switches, and supply appropriate operator's console and labeling. You may wish to color-code the manual push-buttons to identify them from the automatic controls. In this case we need four (4) NO *Start* buttons (Not red), and four (4) NC *Stop* buttons (red), one (1) NC *E-Stop* (Red, Mushroom head), one (1) three position, maintained selector switch, *Hand-Off-Auto* (Figure 7-3), and two (2), two position selector switches for the *Run-Jog* control and *On-Off* switch (Figure 7-4).

Figure 7-3   Three-position HAND-OFF-AUTO selector switch
(Courtesy Allen-Bradley Co., Milwaukee WI)

Figure 7-4 Two-position OFF-ON selector switch
(Courtesy Allen-Bradley Co., Milwaukee WI)

The following are added to our bill of materials. Note that these are not power consuming devices. The minimum number of NO and NC contacts are listed beside each device to help us choose hardware that meets the control requirements.

Control	Qty	NO	NC	Color	Cat. No.
E-Stop	1		1	Red (mushroom)	800T-D6A
Hand-Off-Auto	1	1	1	Black	800T-J2A
Auto Start	1	1		Green	800T-A1A
Auto Stop	1	1		Red	800T-B6A
Stop (1,2,3)	3	1		Red	800T-B6A
Start (1,2,3)	3	1		Orange	800T-A3A
Run-Jog	1	1		Black	800T-H2A
Control Switch	1	1		Black	800T-H2A

All the components shown on the schematic diagram have now been accounted for. For simplicity, the relays have been selected with the minimum contacts required. It is good practice to install relays with the next even number of contacts, to allow for balancing contact spring loading, and provide spares for possible alterations or additions to the circuit.

WHERE WILL THESE ITEMS BE INSTALLED AND WHAT ARE THE INTERCONNECT REQUIREMENTS?

We assume that the power contactors and transformer will be in the motor control center. All push-buttons, selector switches, and lamps will be in an operator's console, and the relays will be in a relay cabinet, possibly mounted beneath the operator's console, or in an adjacent space. This means that we will have three (3) discrete assemblies to wire and inter-connect.

Following each line of logic in Figure 6-4, one must determine where each component is mounted, and <u>terminal numbers</u> that must be provided for inter-connections. Wire numbers, that are inter-connected within an enclosure, usually do not require external terminals. In the following table an (x) indicates that the wire number of a required terminal in the enclosure shown at the head of each column.

Wire No.	Power	Relay	Operator Station	Wire No.	Power	Relay	Operator Station
1	x	x*	x	28		x	x
2	x	x	x	29		x	x
3		x	x	31		x	x
4	x**			32	x	x	
5	x**			33	x	x*	x
6	x	x		34	x	x	
7		x	x	35	x	x*	x
9		x	x	36	x	x	
11		x	x	37	x	x*	x
10		x	x	X2	x	x	x
19		x	x	W1		x	
20		x	x	W2		x	
21		x	x	W3		x	
24		x	x	W4		x	
25		x	x				

* Terminals 1, 33, 35 and 37 have been added to the relay panel as tie points so that a separate conduit run, for the benefit of only four wires, will not be required between the Power source and the Operator's console.
** Wires 4 and 5 are internal OL jumpers between contactors *C1, C2, C3*.

The external wiring scheme is illustrated in Figure 7-5. Note that using only the schematic diagram, we can determine the necessary conduit and inter-wiring requirements for this layout, before the individual components have been assembled. #16 AWG wire would be large enough to handle these low power circuits, but most codes require not less than #14 AWG to meet the mechanical requirements for pulling the wires through conduit. Be sure to check your local code requirements.

Most contractors will add spare wires in each major conduit run, depending upon how many wires that the National Electric Code (NEC) allows in a given conduit size. This gives one some flexibility if a field revision is needed, and also covers any contingency for wires damaged during installation. The labor required to add just one more wire to a lengthy conduit run far exceeds the cost of a few spare control wires, pulled-in during the original installation.

External Wiring Requirements

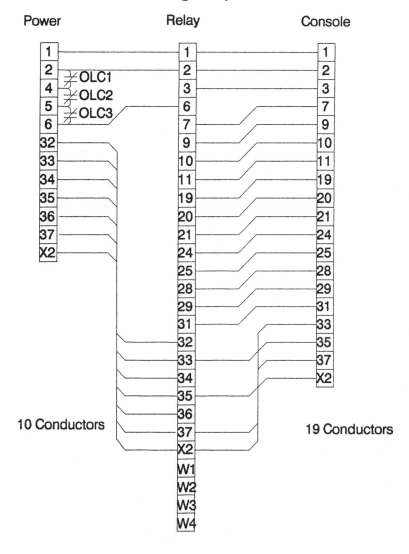

Figure 7-5 External wiring requirements

Chapter 8

Wiring Diagrams

In Chapter 7 we used our schematic diagram to determine the basic bill of materials, and the three areas where the components will be used. Our next project is to develop internal wiring diagrams for the power control center, relay panel and operator's console. Bear in mind that we are working on a fairly "bare-bones" system that could have more indicating lights, interlocked controls to the silo outfeed gate, extra E-stop buttons, and other features.

MOTOR CONTROL CENTER (POWER PANEL)

We are assuming that this installation uses a modular motor control center. This structure has a pre-wired 460v, 3 phase bus system including a horizontal main bus and vertical feeder busses that are accessible for modular "plug-in" power and control assemblies. Motor control modules can have either fused disconnect switches or circuit breaker ahead of the motor contactor. Special modules are available for control transformer modules, relay assemblies, variable speed drives, and special "soft-start" motor controls, to list a few. See photo in Figure 7-1.

Figure 8-1 illustrates a possible configuration for three (3) motor starters, and a control transformer module. Blank areas can receive modules required by other areas of the installation.

Motor starters will normally be pre-wired to incorporate an auxiliary NO holding contact. In this instance, the NO contacts will not be used to seal-in the contactor, but to energize the *green* indicating lamps

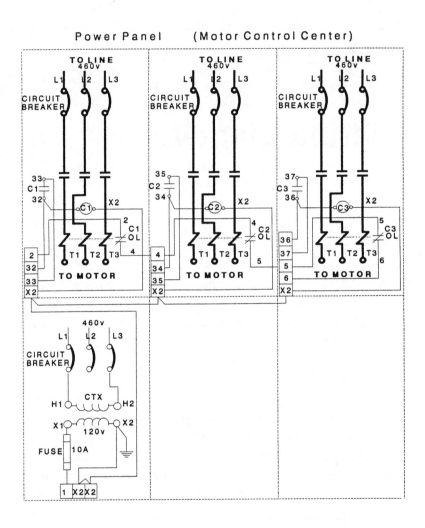

Figure 8-1 Power panel wiring diagram

at the operator's console when the contactor has "pulled-in". This gives the operator proof that a motor starter has actually closed its contacts. We will have to ignore the factory terminal markings, and re-label the contactor wiring to match the schematic drawing. Figure 8-1 shows the required markings for the wires and terminals.

The NC overload contact(s) will originally be factory-wired between the contactor coil and the X2 (low side) of the control circuit for direct protection of each individual contactor. (See Figures 3-1 and 3-2.) Since we are using the three contactors *C1, C2, C3* in a special inter-locked control circuit, we will have to modify this factory overload wiring to meet our special requirements. Return to the schematic drawing in Figure 6-4. On line 10, it shows that *OLC1, OLC2, OLC3* contacts must be connected in series between terminals #2 and #6.

The first step will be to isolate the *OL* contacts from the coil circuits. We need to provide an incoming terminal #2 in the *C1* compartment, and connect this terminal to the NC *C1OL* contact. See Figure 8-1. The other side of this contact becomes terminal #4 and is required in *C2* compartment. Wiring can be run directly to a terminal #4 in *C2* or include an extra outgoing terminal in *C1* (not shown). We repeat this scheme in the *C2* compartment, connecting #4 and #5 across *C2OL*. The circuit then continues in like manner to the *C3* compartment where *C3OL* receives wires #5 and #6. Wire #6 terminates at an outgoing terminal. The actual routing of these wires will be dictated by the structural limitations of the motor control center.

The schematic diagram (Figure 6-4), shows that each coil has independent control from the relay circuits. Lines 28, 30, 32 show the required connections to the *C1, C2, C3* coils. *C1* coil requires terminals #32 and #X2; C2 coil requires #34 and #X2; and C3 requires #36 and #X2. Figure 8-1 shows terminals required to accomplish this.

Since the control power is derived from a control transformer, located in the same motor control center, we have shown it as a separate module, with outgoing terminals #1 and #X2. An extra #X2 jumpered terminal is shown in this compartment to allow multiple connections without overcrowding the terminals with more than *two (2) wires per side of a termination.* Remember that #X2 will not only be required for the motor contactors, but also for relays, and indicating lamps in their respective locations. Most jurisdictions require that #X2 must be bonded at the local ground bus to limit the maximum potential to ground to approximately 120 vac.

A total of ten (10) outgoing wires will be required for connections to the relay panel, as shown in Figure 7-5.

RELAY PANEL

The relay panel requires some creative planning. From the "Bill of Materials" and catalog information, we can determine the minimum area required to mount the relays, and thus the size of the enclosure. It is often expedient to make full size templates of the relays, and use them as "paper dolls" to work up the panel layout. If possible, the relays should be laid out in a functional pattern so one can watch them function in their logical sequence.

Figure 8-2 shows a typical layout. Note that in the second, third and fourth horizontal line of relays, that relays in each level are related to the same conveyor control. Conveyor C1 is controlled by ST1, A1, M1; Conveyor C2 is controlled by ST2, AT2, M2; and Conveyor C3 is controlled by ST3, AT3, M3. The initiating timer T, Overload relay OLR, and warning relay W are installed at the top. The jog relay JR, is shown alone, and is the relay with maximum number of contacts.

It may be impossible to show the contacts in their actual physical position on your drawing, but they must be labeled functionally to identify instantaneous contacts as contrasted to timed contacts on the same device. The author always includes the letter "T" in the nomenclature of any relay that has a timing function. Many assemblers specify a minimum of four contacts per relay to provide for possible revisions, and spares. Figure 8-2 shows only the required contacts to simplify the drawing.

Once the lay-out and contact arrangement is drawn into the drawing, we add our terminal strip for incoming and outgoing connections per Figure 7-5, and proceed with the internal wiring plan. Most assemblers now use plastic wiring gutters in a pattern between the relays to keep the wire bundles neat, and readily accessible.

The actual wiring can be shown "point to point" as shown in many connections in Figure 8-2, (see #X2, and others), but it is obvious that the drawing will become so cluttered with lines that it will be very hard

Relay Panel

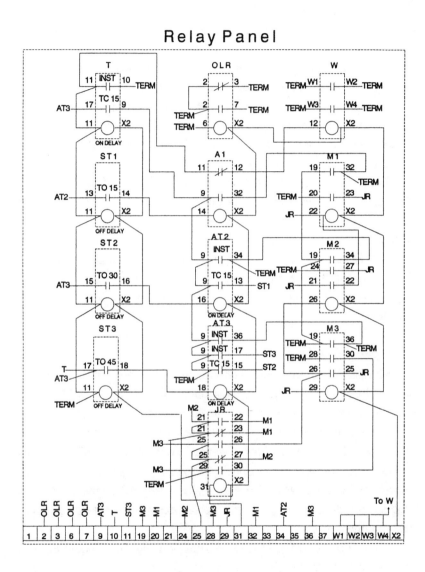

Figure 8-2 Relay panel wiring diagram

to follow. The alternative is to use a "destination address" system as illustrated at most terminal locations. If you follow wire #19 from the terminal strip, you will see that it is addressed to *M3*. When you locate #19 on *M3*, it is addressed back to *TERM*, and also continues to *M2* and *M1*. This gives the wireman an easy way of following a wire number without becoming lost in a maze of crossing lines.

Terminals #1, #33, #35 and #37 have no internal connections, but are provided for interconnection between the power panel and operator's console. Terminal #X2 can be doubled, if necessary, to allow for extra connections.

Timers have been labeled for "on-delay" and "off-delay" service. The catalog numbers that were originally selected in Chapter 7 identify universal time relays that are convertible to either duty, and also have convertible NC or NO contacts.

In this discussion, we have not identified special Underwriters Laboratories (UL), or other agency, requirements to meet local code requirements. Meeting those requirements is the responsibility of the assembler.

CONTROL CONSOLE

All operator controls and indicating lamps are installed in this assembly. Again, we always try to make a functional lay-out that will give the operator an intuitive sense of how things are related. See Figure 8-3.

The top row of devices include the *Control Power* switch, *Auto-Start, Auto-Stop* and the *Red Overload* warning lamp. The left side includes the mushroom head *Red Emergency Stop* push-button, and selector switches for *Hand-off-Auto* and *Run/Jog*. The manual *Run* and *Stop* push-buttons are aligned with the respective *Green* lamps that indicate running conveyors.

Point to point wiring is shown in this drawing, but it could be "destination addressed" as illustrated in the relay panel. Since it has to be wired from the rear, a mirror image drawing is sometimes produced for this assembly to make wiring easier.

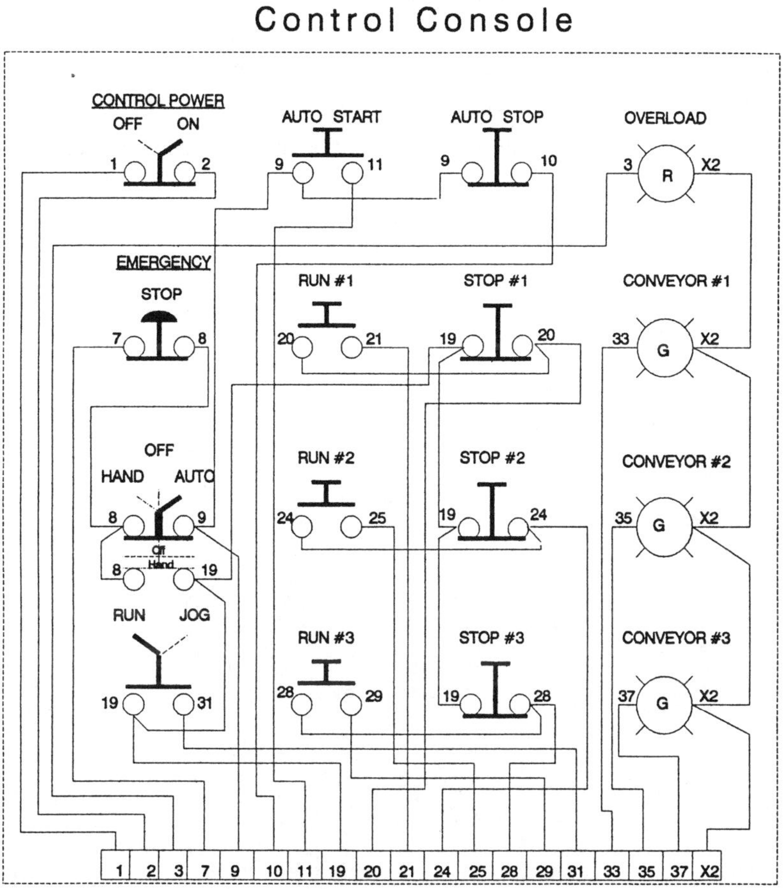

Control Console

Figure 8-3 Control console wiring diagram

SUMMARY

Beginning with a written description of an application in Chapter 4, we developed the required logic and converted it into an electrical circuit. Then this information was used to determine the required components and system layout, and finally the wiring diagrams for each major assembly. The entire process is an extension of logic, building continuously upon each previous step in the process. In the next chapter, we will see how these drawings can be used for expeditious trouble-shooting.

Chapter 9

Application of Schematic and Wiring Diagrams

Assume that several weeks, or months, have passed since you prepared the drawings in the previous chapters. You have been pre-occupied with other projects since you shipped your equipment to your customer, and probably do not remember all the details of this project. His installation is now completed, and you receive a call to provide start-up services at the job site. Wiring has been installed by others, so you will have to inspect the installation to be sure that it is safe to operate. This procedure is often called "commissioning".

Using the wiring diagrams, you will locate each assembly at the facility and check for proper connections at each panel and console area with power off. This preliminary check might include a wire count at each terminal strip and a close inspection of the motor control center to see that proper modifications were made for combining the overloads, isolating the *C1, C2, C3* coil circuits, and re-labeling the auxiliary contacts.

The installer or an operating electrician should accompany you during all start-up operations so that he will understand your system when you hand it over to him. He can also assist in making any corrections that are needed. In some jurisdictions, you may be required to use a local union electrician to make revisions, under your supervision.

CONTROL TESTING

Since this installation has three (3) motors that are operating at remote locations, we want to assure that our logic is correct before any motor starts. An improper start might damage equipment or endanger other workmen. The first thing we must do is to ascertain that the motor circuit breakers are <u>locked</u> and tagged in the *OFF* position. This will allow operation of the contactors from the control circuits without starting any motors. The circuit breaker for the control transformer is now turned *ON*.

Our schematic, Figure 9-1 (a copy of Figure 6-4), is the all-important document for the following tests. At this time, we will use our voltmeter to test for 115 volts across terminals #1 and #X2 at the transformer module. We will also check for the same reading between #1 and ground. This proves that our secondary fuse is not blown, and control power should be available at the relay panel and operator's console.

The next item in the schematic is the *On/Off* selector switch at the operator's console, so we proceed to that location. Again, we test for 115 volts across terminals #1 and #X2 to prove the wiring from the power source. If voltage is correct, we also know that terminal #1 in the relay panel is correctly wired.

After proving that our control power is available, we can now turn ON the console power control switch, and check for voltage across #2 and #X2. At this time your helper should check the relay panel to see that *OLR* is energized. If he momentarily disconnects a wire from terminal #6, with power ON, *OLR* should drop out and the overload indicating lamp in the operator's console should come ON. At this point we have proven that there is continuity through the overload relay NC contacts, and the overload lamp will indicate whenever *OLR* trips OFF. Now, with all wires connected, test for voltage across #7 and #X2. If 115 volts is shown, we have proven our schematic to the Emergency Stop push-button on line <u>11</u>.

MANUAL STARTUP

Our next test is to check for voltage across the *Hand/Off/Auto* selector switch. Switch to *Hand* and test for voltage between terminals #19

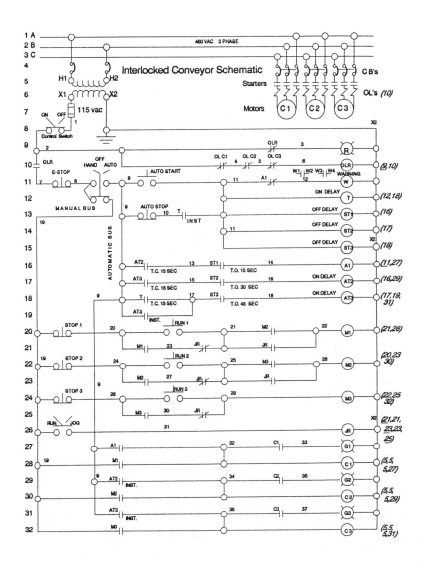

Figure 9-1 Conveyor schematic diagram

and #X2 at the relay panel. Then switch to *Auto* and test for voltage between terminals #9 and #X2. See Figure 8-3. If you press the *Emergency Stop* push-button during these tests, it should always interrupt the power.

We are now ready to make some operational tests. For safety reasons, it is a common practice to test individual *Hand* operations first, so we will turn our selector switch to *Hand*. Then select *Jog*, at which time the *JR* relay should pull in. Using the *Run* buttons, check that the corresponding *green* indicator lamp comes ON when each is pressed in sequence and go out when the button is released. Your helper should go to the motor control panel at this time and observe that the proper contactor is being energized at the same time.

Problem! C2 is energized when *Run 1* is pressed, and *C1* is energized when *Run 2* is pressed. The *green* indicating lamps have the same incorrect reading, Where is the problem?

Looking at our schematic, lines 28 and 30, we find that terminals #32 and #34 terminate at *C1* and *C2* respectively. Since our lamps match the contactor actions, we can assume that wiring has been interchanged somewhere in the *C1* and *C2* controls. Locate #32 and #34 at the relay panel, and test each for voltage to #X2 as each respective button is pressed. We find that #32 and #34 signals have been interchanged between these two terminals.

Next, we have someone observe relays *M1* and *M2* in the relay panel as we press *Run 1* and *Run 2* respectively. We find that the relay action is crossed so the wrong one is being energized.

We now make a voltmeter test at terminals #21 and #25 in the relay panel, and find they are apparently crossed. Wire numbers appear correct, so we then go to the operator's console. Upon close inspection, we find that the wires, bundled between the terminal strip and the two push-buttons, were mislabeled at the push-button connections, creating a crossed connection. We exchange connections at the terminal strip, and correct the labels on the two wires. Re-testing shows that the proper contactors are pulling in. (Note that there may have been other faults that could have created this problem).

We will now turn our *Run/Jog* selector switch to the *Run* position. If we press *Run 1* first, nothing happens. Likewise, pressing *Run 2* has no effect. When we momentarily press *Run 3*, the Conveyor #3 indicating lamp G3 will turn ON and stay lit. If we press the *Stop 3*, the lamp will go out. Further testing proves that you can start and run all conveyors if they are energized in the order of 3-2-1. With all running, press *Stop 3*. All conveyors should stop. If all are running, and you press *Stop 2*, conveyors #1 and #2 will stop, and #3 will continue to run. Pressing *Stop 1* will stop #1 conveyor, and the others will continue to run. We have now proven the manual mode control logic.

AUTOMATIC STARTUP

Stop all conveyors and switch the selector switch to *Auto*.

Press *Auto Start* momentarily. The relays *W, T, ST1, ST2, ST3* should energize immediately, and stay ON. After 15 seconds, *AT3* should be energized, and *G3* indicating lamp should come ON. At 30 seconds, *AT2* should be energized, and *G2* indicating should turn ON. At 45 seconds, *AI* should be energized, *G1* indicating lamp should turn ON, and warning relay *W* should turn OFF.

Problem! Conveyor #1 does not start, and warning does not turn OFF.

Refer to Figure 9-1. On line 27 we see that *AI* must be energized to turn *C1* and *G1* ON. On line 11 we note that W will be turned off when *AI* is energized. This tells us that *AI* is not turning ON as expected. On line 16 we see that *AI* requires contacts *AT2* and *ST1* closed to connect it to the hot line #9. Our contact listing on the right margin of the drawing indicates that *AT2* has as contact on line 29 that is used to turn on *C2* and *G2*. Since *G2* is energized, we know that *AT2* is being energized.

We now consult our wiring diagram, Figure 8-3, and locate *ST1*. Upon inspection, we find that *ST1* is energized. We can use our voltmeter to test between coil connection #14 on *AI* and contact connections #13 on *AT2* and *ST1*. Our voltmeter shows no voltage at *ST1* #13, but does read at *AT2* #13. This means that there is an open circuit between these two terminals. Upon inspection, we find that the jumper

#13 was poorly stripped at *AT2* during assembly, and was clamped on the wire insulation, instead of the bared wire.

We turn our selector switch to *Off* while we repair the jumper #13, then return it to *Auto* and try *Auto Start* again. This time the system cycles properly.

After a successful automatic start cycle, we now have to prove the automatic stop performance. With the system in *Auto Start* mode, and all conveyor lamps lit, we now momentarily press the *Auto Stop* push-button. We expect relays *T, ST1, ST2, ST3* to drop out immediately. After 15 seconds, *A1* and *C1* drop out, and the *Conveyor #1* indicating lamp *G1* should go out. At 30 seconds, *AT2* and *C2* drop out and turns Off the *G2* indicating lamp. At 45 seconds, *AT3* and *C3* drop out and turn Off *G3* indicating lamp.

POWER TESTING

There is one more test we must conduct. The motor rotation must be proven for each conveyor. Running the motors in reverse may damage belt tightening hardware, so we dare not allow much movement in reverse. To make this test, we want maximum control of the conveyors, so we will set our controls for *Hand* and *Jog*.

Each motor will be tested separately by turning its circuit breaker ON, and then observing the motor rotation while momentarily jogging each motor with the respective *Run* push-button. If a motor starts in reverse, the circuit breaker is turned OFF and two motor leads are interchanged at the "T" terminals on the contactor. Then turn it ON and re-test.

The conveyors are each about 100 feet long, and designed to run at 400 feet per minute. This means that a conveyor should totally load or unload in 15 seconds. If the conveyor dimensions turn out to be shorter, or longer, we may have to "fine-tune" our timers by a few seconds to insure that all belts are cleared during startup, and leave no material on the belts during a normal automatic stop cycle. A slight variance in gear ratios would also require tuning.

This start-up and trouble-shooting example has shown the importance of understanding the logic from a schematic view of the operation, and applying this information to the separate wiring diagrams to locate the actual component that may be at fault. If you had tried to solve the problems, using the wiring diagrams, it may have taken hours of wire-by-wire tracing, through three (3) wiring diagrams. The schematic "ladder-logic" lets you quickly pin-point the probable fault point, then go to the wiring diagram to find and fix it.

One should remember *safety* is the most important consideration during start-up tests, and nothing should be taken for granted. We cannot always assume that the installing electrician has "already" proven a circuit that you are unsure of, in spite of his claims. He may have a completely different idea of how the logic works, and honestly believe that his tests were adequate, when the opposite is true. The author has been faced with this many times, and it does require a lot of diplomacy at times to clear a problem, especially when working on foreign installations with a language barrier.

All possible tests should be made with power removed from motors and critical actuators. Someone should always be near the *Emergency Stop* button when power is applied to interrupt all action in the event of a fault. A damaged conductor may require total isolation from power so it can be tested with an ohmmeter or megger for a ground, short or open circuit. One's understanding the schematic ladder diagram is most important when trouble-shooting complex installations.

Chapter 10

Case History Example

In the preceding chapters, we developed a schematic for the sequential control of conveyors. We will now look at a relatively simple application that includes *positional* control. This is really a "case history" example, based upon experience.

In the early 1950's, the author operated a small electrical contracting business in southern Oregon. With arrival of REA power lines, and a "boom" in logging, a number of lumber mills were built to process the timber locally. Mills were not large enough to require a full-time plant electrician, so most construction and maintenance was provided by the local contractor.

One of the more progressive sawmills decided to add a planer mill to get into the finished lumber market. The local millwrights fabricated all lumber handling equipment on-site, and the electrician had to build controls to match whatever they dreamed up.

Figure 10-1 shows part of the feed system to the planer mill where bundles of stacked, unfinished lumber are broken-down before the lumber reaches the planer. The following describes the operating sequence.

A fork-lift operator delivers the bundles to an infeed chain conveyor. The conveyor moves the bundle over the forks of a "tilt-hoist" assembly. We want this conveyor to load the forks automatically whenever a bundle is available, and the forks are in "home" position. A manual control has to be provided to handle any emergency conditions.

Tilt/Hoist

For Lumber bundle breakdown

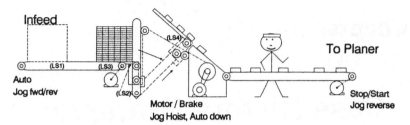

Forklift places bundle on infeed
If Tilt/Hoist is empty, bundle moves forward and loads elevator forks.
As operator uses foot switch, the winch tilts the hoist against stop.
When operator continues to jog his foot switch, the hoist lifts.
As each tier of lumber reaches the top, it slides off the bundle.
When hoist is empty, operator presses "hoist down" button.
Forks return to bottom, and hoist tilts to verticle
LS1   bundle at infeed
LS2   bundle is on forks
LS3   elevator at max down "home" position
LS4   elevator at max up position
Operator controls---
       Control power on/off
       Infeed hand/auto
       Infeed jog forward
       Infeed jog reverse
       Hoist jog up footswitch
       Hoist down
       Outfeed start
       Outfeed stop
       Outfeed jog reverse

Figure 10-1   Tilt/hoist plan

When the operator actuates the tilt-hoist with his foot switch, it will first tilt over against a bumper, and then, as the winch continues to run, the bundle will be lifted up. By jogging this *Hoist* foot switch, the operator can unload one tier of lumber at a time as it clears the top of the bumper and slides down onto the inspection table chains.

To reverse the hoist, the operator will press a *Down* push-button, and it will automatically return to "home" position and stop. As the tilt-hoist drive is reversed, the forks return to the lowest position, and then over-balance the "tilt" mechanism, returning it to a vertical position. Automatic operation is required because the operator not only inspects and straightens the lumber, but is also responsible for keeping the lumber moving smoothly through the planer where the boards are fed end-to-end at about 800 feet per minute.

When we arrived on the scene, motors and motor/brake were in place, but the only instructions we had were, " – make it work". Now we have to exercise our logic!

MOTIONS REQUIRED

We decided that all three motors needed reversing control. There is always the possibility of a bundle fall-down if the fork lift operator is careless. A *Jog Reverse* would help back the load out to help clean-up. We already know that the hoist drive must be reversible. The inspection table motor usually runs continuously, but is susceptible to jam-ups where the lumber changes direction as it enters the planer infeed. A *Jog Reverse* would be necessary to pull the boards out of the jam-up. We now know that we will be using three (3) reversing contactors.

POSITION DETECTION

To operate anything in an automatic mode, one needs to know the *position* of all components that affect the operation. In this case, we need to know where the lumber bundle is, and the position of the tilt-hoist forks.

We do not want the infeed to run unless it has a load on it, so we will install a limit switch LS1 between the infeed chains at the location where the fork-lift delivers the load to indicate that a load is present. It is mounted so that the load depresses the limit switch actuator when present.

The infeed must stop whenever the load has reached the end of travel over the forks. We will mount limit switch *LS2* on the infeed conveyor in like manner at a position where it will be tripped as the load arrives at end of travel.

The bundle must not move forward if the forks are not in the "home" position. This requires a limit switch *LS3* mounted on the infeed conveyor at a point where the forks will provide a signal when the forks are horizontal, and below the infeed chain level.

A fourth limit switch *LS4* is required on the fork carriage to provide a maximum height shut-off when the forks reach a point where the last tier of lumber unloads. Since the very busy operator uses a foot-switch to jog the elevator in the up direction, he may not always realize when it has reached its limit.

POWER SYSTEM

The mill operates on 460 volt, three-phase, 60 Hz power. Because of the moisture that is present in this coastal climate, it is obvious that we should reduce our control power to 115 volt, single phase. We can combine a control transformer in the same enclosure with our contactors and any relays that may be needed.

OPERATOR CONTROLS

The operator station will require a minimum of the following control devices.

— Control power On/Off	— Hoist Down
— Infeed Hand/Auto	— Outfeed Start
— Infeed Jog Forward	— Outfeed Stop
— Infeed Jog Reverse	— Outfeed Jog Reverse
— Hoist Foot Switch (jog-up)	

SCHEMATIC DIAGRAM

Figure 10-2 shows a schematic diagram that will provide the logic we need. (We hope that it is easier to read than the one we developed on that mill floor, with a piece of wrapping paper, over 40 years ago.)

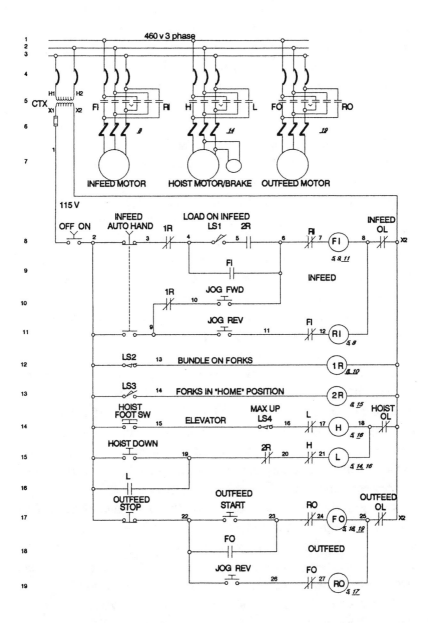

Figure 10-2 Tilt/hoist schematic wiring diagram

We have shown a typical power schematic on lines 1 through 7, including motor circuit breakers, contactors, control transformer, and the respective motors. You will see on lines 12 and 13 that limit switches *LS2* and *LS3* are connected to relays to provide additional contacts. This is simpler than running extra wires out to multiple contacts in the limit switches.

On line 8 shows the control power *On/Off* switch and the *Infeed Auto/Hand* selector switch. Continuing across this line is the rung that automatically moves a bundle forward on the infeed conveyor. If the selector switch is set on *Auto* and *LS1* has a load on it, and *2R(LS3)* is energized to prove that the forks are in the "home" position, *FI* will energize and seal in through its contact on line 9. The conveyor will continue to run until *1R(LS2)* opens as the load reaches the end of travel.

Starting on line 11, if the selector switch is on *Hand*, the infeed can be jogged in the forward and reverse directions, with *1R* protecting the forward travel limits. *FI* and *RI* are interlocked by their respective NC contacts to prevent simultaneous operation.

LS2 on line 12 is NC, and only opens when a bundle has reached the end of travel over the forks. *LS3* on line 13 is NO and is never closed unless the forks are in position to receive a load.

Line 14 shows the foot-switch circuit that provides *Jog Up* control for the hoist. *LS4* is NC, and opens if the hoist has reached its upper limit, to prevent further movement in that direction. If the operator momentarily presses *Hoist Down* on line 15, the lower contactor *L* is energized, seals itself on line 16, and continues to run until the forks have reached "home" position as indicated by *2R*. *H* and *L* are interlocked by their respective NC contacts to prevent simultaneous operation.

Line 17 and beyond controls the forward/reverse control of the outfeed conveyors through *FO* and *RO*. *FO* is controlled by the typical start/stop circuit, while *RO* has a simple jog/reverse circuit. Typical NC interlocks are shown to prevent simultaneous operations. Note that all contactors have individual overload protection as provided from the factory.

In this type of control it is important to carefully place the limit-switches, because there is real danger of breaking something if limits are exceeded. One can imagine the mess one could make if the infeed conveyor shoved a bundle forward when the elevator forks were in a tilted position. This is a fairly simple circuit to design and install in the field, but you should also document it with follow-up drawings.

WIRING DIAGRAMS

Figure 10-3 shows a field wiring diagram derived from the schematic. Most integrated motor-brake assemblies are designed to release the brake whenever the motor is energized. The brake wiring is usually connected directly to the "T" leads of the motor, whether single or three phase. The brake voltage should always be verified before connections are made, because there are times when special voltages, AC or DC may be required for the brake.

Figure 10-4 illustrates a simplified wiring diagram for the power and control panel. It includes the control transformer, relays, and reversing contactors. Contactors are shown with typical factory wiring arrangements, modified to suit our schematic requirements.

Figure 10-5 shows the operator's console wiring. Note that the controls are aligned in order as the material flows. If necessary, the order can be reversed to accommodate which way the operator is normally facing.

SUMMARY

In previous discussions we have emphasized the fact that every rung of logic either stores information, or performs a specific task. In this example, limit switches have been used to tell us if something is at a particular location, or is not there. The other "action" rungs are enabled or disabled by these "position" indicators. The circuit even prevents the operator from over-riding the control and causing problems.

In the trade this is often referred to as making the system "idiot-proof." It is amazing how often you find someone wandering around, pushing buttons, with no idea of the damage he can cause. You always want to test your logic to be sure that there is no way to get your system out of sequence, using the provided controls.

In some cases redundant limit switches are required to cover the possibility that one might fail. If NC limit switches are connected in parallel, *both* must be operated before the circuit changes state. If connected in series, *either* will open the circuit.

On the other hand, if NO contacts are connected in series, *both* must be operated before the circuit changes state. If connected in parallel, *either* will energize the circuit.

The chapter that follows illustrates several unique tricks and logic patterns that can be used for special requirements. There really is no limit to how logic can be applied to provide solutions.

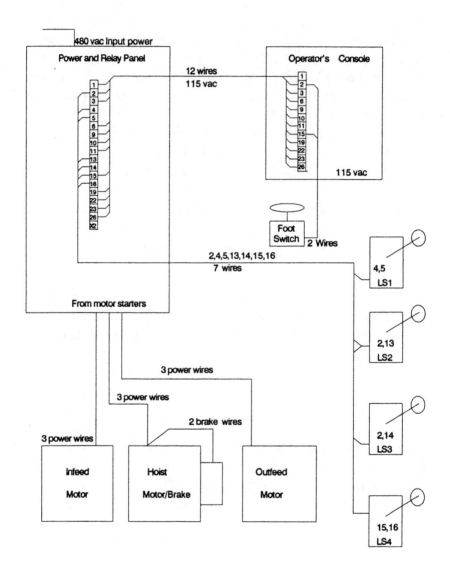

Figure 10-3 Tilt/hoist wiring schedule

POWER AND CONTROL PANEL

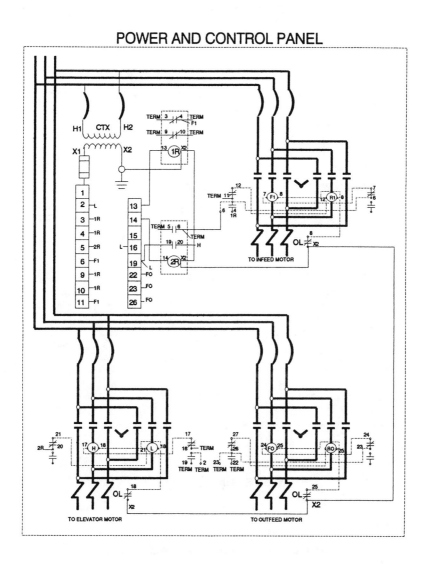

Figure 10-4 Tilt/hoist power and control panel wiring diagram

Operator's console

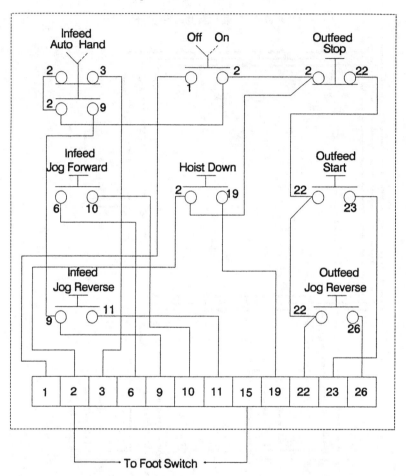

Figure 10-5 Operator's console wiring diagram

Chapter 11

Special Examples of Relay Logic

We have shown how to logically build a ladder diagram, and develop the necessary companion drawings to complete an installation. There are times when unusual functions are needed that may require some special design ingenuity. In the absence of a programmable logic controller (PLC), we can still use relays to perform many interesting functions. The following examples illustrate basic logic that may be modified as necessary to fit your requirements. You will find that there are many variations of these circuits in industry – in other words, there are other circuits that accomplish the same functions.

BAILING CIRCUIT

Figure 11-1 illustrates one way to assemble a *bailing circuit*. We have a requirement where we are selecting three functions controlled by relays *A,B,C*. They are selected by momentarily pressing push-buttons *1,2,3* respectively. The last selection determines which relay is selected, and automatically cancels any previous selection. It is impossible to seal-in more than one relay.

This simple circuit can be accomplished by substituting NC Push-button contacts for the NC relay contacts. In this case *#1* push-button would have two additional NC contact blocks replacing NC A contacts, etc.

BAILING CIRCUIT

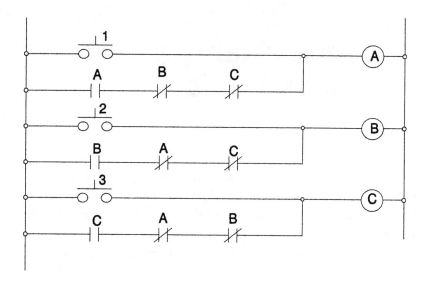

Figure 11-1   Bailing circuit

A SORTING APPLICATION

Consider an application where product must be sorted into different bins. As product passes the inspection point, the operator determines which bin it will be diverted into. This application requires three (3) unique relay applications, *encode, shift-register,* and *decode.*

The operator presses his selector button, as the package passes through a *read* zone, to assign the destination number *(encode)*. A *shift-register* circuit moves the number through the several diversion zones as the product moves down the conveyor. Each zone reads the number *(decode)* as it passes, diverting the package if it does match or ignoring it if it does not match. We will show samples of each type of application, but will make no attempt to show a completely integrated system. These examples can be used as sample "building blocks" to develop your own unique control system.

BINARY LOGIC

Relays can be used to set-up numerical codes by using all the possible contact combinations that can be arranged from a group of relays. For instance, there are eight (8) different combinations that can be made, using only three (3) relays.

An easy way to keep track of the combinations is to use *binary* logic. A relay has only <u>two</u> conditions, – On or Off. When we write decimal numbers, the right column has a value of one (1) times the number (10^0), the second column has a value of ten (10) times the number (10^1), and the third column value is one hundred (100) times the number (10^2).

Binary numbers have only zeroes (0) and ones (1). The right column has a "weight" of one (1) or (2^0), the next column two (2) or (2^1), and the third column four (4) or (2^2). If we add the values in all columns, we have the equivalent decimal number. Using three relays, we assign a values of 1,2 and 4 to each respective relay.

The following table shows the decimal equivalent of each binary value.

Binary	Decimal
000	0
001	1
010	2
011	3
100	4
101	5
110	6
111	7

See Chapter 13 for further discussion of number systems.

ENCODING

One method of encoding is shown in Figure 11-2. Zero (0) is a real option, so there are a total of eight (8) possible conditions. In this case we have shown a diode logic circuit for momentary encoding. Note that push-buttons *1,2,4* can only energize relays 1,2,4 respectively. PB *3* energizes relays 1 and 2, adding up to a total of three (3) – binary 011. PB *7* energizes relays 1,2 and 4 for a total of seven (7) – binary 111. The PB *0* is used through relay 0 to correct (cancel) any code in the "read" zone, and in effect install zero as the code in that zone of a shift-register (Figure 11-3).

SHIFT-REGISTER

A shift-register is logic circuit that can move encoded information sequentially through a series of zones. Different data can appear in adjacent zones as it is moved forward without loss or interference. Data can be "read" as the information passes through a zone, and acted upon if the code matches that zone's specific code, or ignored if there is no match.

READ CODE INTO SHIFT-REGISTER

The next step is to capture the code into the shift-register. The shift-register is divided into several zones. For our illustration, zone A is our "read" zone, zone B is a free zone to carry the code forward to the first "unload" zone. Zones C through J provide eight opportunities to selec-

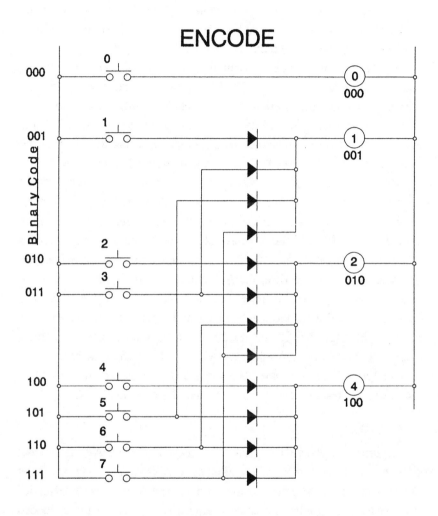

Figure 11-2   Diode encoding circuit

tively unload product. As a rule, the highest number is used in the first unload zone, because as the binary code becomes 011 (3) or less, we do not have to carry a #4 relay, and eventually #2 and #1 relays, to the end of the register, saving wiring and equipment. This is illustrated in Figure 11-5 at the end of this section.

Figure 11-3 illustrates a circuit that will read the code. If relays 1 and 2 were energized by PB *3* in Figure 11-2, as the product actuates the "Enter Code" limit switch, relays *1A* and *2A* will seal-in. Relay *4A* will remain de-energized. We have now locked the binary code 011 into zone A. If a *0* (zero) is entered as a correction during this interval, all "read" relays will be turned off and the zone will effectively read "zero".

Zone B is interleaved in this drawing to show how the code can be moved from the "read" zone A, to zone B, then reset zone A to zero. Look at relay *1A* as an example. When it seals-in during code entry, it also closes a contact ahead of relay *1B* in zone B. As the product moves into zone B, it closes the "shift" limit switch which can now energize and seal-in *1B*. At that time, the NC contact *1B* opens the circuit to *1A* and it is reset to zero. In other words, we have moved our code to a new zone, sealed it in, and reached back and turned off zone A, to allow entry of the next code. Since we did not have a code calling for relay *4A*, it did not seal-in, nor did it pass any signal into the next zone.

Relays *1B, 2B, 4B* have NC contacts from *1C, 2C, 4C* ahead of their coils for reset purposes when the signal arrives in zone C.

DECODING

Figure 11-4 shows two "unload" zones, C and D. As the product proceeds from zone to zone, it continues to advance the code, and reset the previous zone to zero. As the product passes over its "read code" limit switch, it samples the contact pattern set up by the binary coded relays. Since we have arbitrarily used the binary code 011 for our example, *4C* would not be energized, and zone C would reject the code, and allow the product to pass by. The same would be true for zone D.

If we send a binary code of 110, zone D will recognize it, energize *D* to set-up an unload timer *UDT,* and reset its own binary relays to zero. The timer is expected to initiate whatever sequence that is required

Code to Shift Register

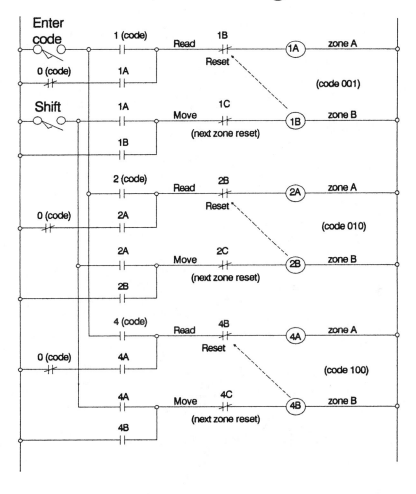

Figure 11-3 Code to shift-register circuit

Shift to Zones C & D and Decode

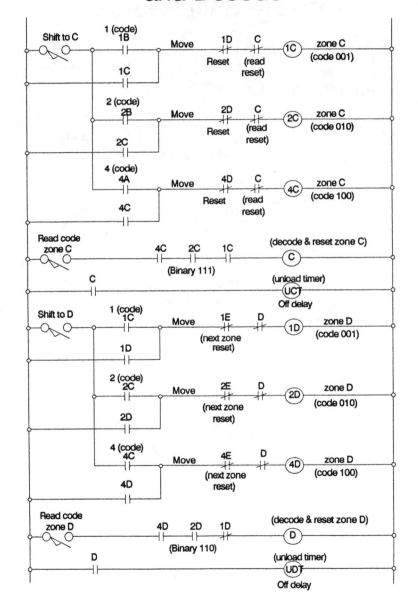

Figure 11-4 Two zones of a shift-register circuit

to unload the product. Other zones (not shown) would be constructed along this same pattern.

By adding one more binary relay per zone, we could increase our codes to handle sixteen (16) combinations – decimal 0 through 15.

Figure 11-5 shows our decode pattern for the example. As mentioned above, when we reach a zone that has a code less than binary 100 (decimal 4) we do not have to carry that relay in our pattern. So zones G through J can omit #4 binary relays. Likewise, zones I through J do not need #2 binary relays. Zone J needs no decoding relays as it can default to zero, and unload anything that reaches it.

COUNTING WITH RELAYS

In our previous example, we might assume that one or more of the unload zones requires that when three items have accumulated in an unload zone, they would be moved to a packaging area. In other words, we want to move the product along in groups of three.

Figure 11-6 shows a simple counting relay circuit. The push-buttons could be replaced by strategically located limit switches or other relay contacts.

When "reset" is actuated, relays $1T$, $2T$, 3 are energized and sealed-in. The "count pulse" energizes C each time the push-button is pressed. At the first momentary pulse, $1T$ is turned off. A $1T$ contact bypasses the C contact ahead of $2T$ during the pulse, to prevent $2T$ from dropping out, then opens after a time-delay.

At the second momentary pulse, $2T$ will drop-out, and then release the by-pass on relay 3. On the third pulse 3 will drop-out and energize the action timer AT.

The time-delay relays must be adjusted so they do not time-out before the count pulse is released. The AT timer could be used to energize relay R for automatic reset, if desired. This is a relatively simple circuit, so there are limitless variations that can be devised.

It is obvious that programmable logic controllers (PLCs) can provide these functions much easier than using relays. However, if you do not have access to a PLC you can still improvise with relays, to accomplish a lot of interesting control logic.

Typical Decoding Pattern

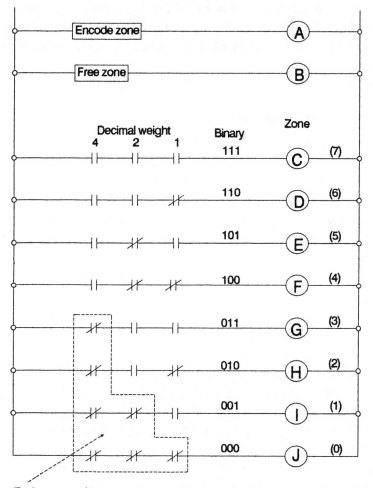

Relays and contacts can be omitted to save hardware

Figure 11-5 Typical decoding pattern

Counting with relays

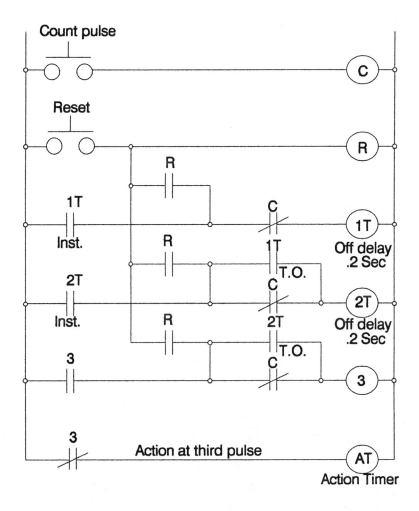

Figure 11-6 Counting with relays

Chapter 12

Introduction to Programmable Logic Controllers

The development of Programmable Logic Controllers (PLCs) was driven primarily by the requirements of automobile manufacturers who constantly changed their production line control systems to accommodate their new car models. In the past, this required extensive rewiring of banks of relays – a very expensive procedure. In the 1970s, with the emergence of solid-state electronic logic devices, several auto companies challenged control manufacturers to develop a means of changing control logic without the need to totally rewire the system. The Programmable Logic Controller (PLC) evolved from this requirement. (PLC™ is a registered trade mark of the Allen-Bradley Co. but is now widely used as a generic term for programmable controllers.) A number of companies responded with various versions of this type of control.

The PLCs are designed to be relatively "user-friendly" so that electricians can easily make the transition from all-relay control to electronic systems. They give users the capability of displaying and trouble-shooting ladder logic on a cathode ray tube (CRT) that showed the logic in real time. The logic can be "rewired" (programmed) on the CRT screen, and tested, without the need to assemble and rewire banks of relays.

The existing push-buttons, limit switches, and other command components continue to be used, and become *input* devices to the PLC. In like manner, the contactors, auxiliary relays, solenoids, indicating lamps, etc., become *output* devices controlled by the PLC. The *ladder logic* is contained as *software* (memory) in the PLC, replacing the inter-wiring previously required between the banks of relays. If one understands the interface between the *hardware* and the *software,* the transition to PLCs is relatively easy to accomplish. This approach to control allows "laymen" to use the control without necessarily being expert computer programmers.

The following introduction to PLCs should be considered generic in content. The author programmed at least five different brands of PLCs, under contract with original equipment manufacturers (OEMs), when they were required by their customers to convert their machines from relays to PLC control.

While each PLC manufacturer may have unique addressing systems, or varying instruction sets, you will find that the similarities will outnumber the differences. A typical program appears on the CRT as a ladder diagram, with contacts, coils, and circuit branching, very similar to that which appears on an equivalent schematic for relay logic.

Our intent is to help one understand the transition from relays to PLC control, rather than trying to teach the details of designing and programming a specific brand of equipment. Manufacturers have numerous programming schools, from basic to advanced training, and you should consider attending them if you plan to become a proficient programmer.

PLC HARDWARE

Programmable controllers have a modular construction. They require a *power supply, control processor unit (CPU), input/output rack (I/O),* and assorted *input* and *output* modules. Systems range in size from a compact "shoe-box" design with limited memory and I/O points, as shown in Figure 12-1, to systems that can handle thousands of I/O, and multiple, inter-connected CPUs. A separate *programming device* is required, which is usually an industrial computer terminal, a personal computer, or a dedicated programmer.

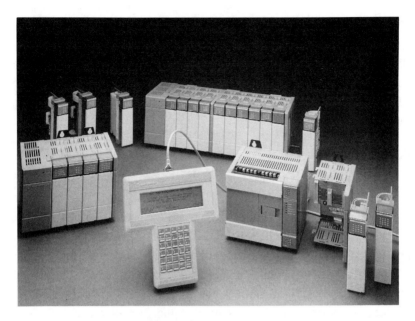

Figure 12-1 Small logic controller, SLC 500, showing variety of modules (Courtesy Allen-Bradley Co., Milwaukee WI)

Power Supply

The internal logic and communication circuitry usually operates on 5 and 15 volt DC power. The power supply provides filtering and isolation of the low voltage power from the AC power line. Power supply assemblies may be separate modules, or in some cases, plug-in modules in the I/O racks. Separate control transformers are often used to isolate inputs and CPU from output devices. The purpose is to isolate this sensitive circuitry from transient disturbances produced by any highly inductive output devices.

CPU

This unit contains the "brains" of the PLC. It is often referred to as a *microprocessor* or *sequencer.* The basic instruction set is a high level program, installed in Read Only Memory (ROM). The programmed logic is usually stored in Electrically Erasable Permanent Read Only

Memory (EEPROM). The CPU will save everything in memory, even after a power loss. Since it is "electrically erasable", the logic can be edited or changed as the need arises. The programming device is connected to the CPU whenever the operator needs to monitor, troubleshoot, edit, or program the system, but is not required during the normal running operations.

I/O rack

This assembly contains slots to receive various input and output modules. The rack can be local, combined with the CPU and power supply, or remote. Each rack is given a unique address so that the CPU can recognize it. Within each rack, the slots have unique addresses. Power and communication cables are required for remote installations. The replaceable *I/O* modules plug into a back-plane that communicates directly with the CPU or through the cable assembly. Field wiring terminates on "swing arms" that plug into the face of the *I/O* modules. This allows a quick change of *I/O* modules without disconnecting the field wiring. Every module terminal also has a unique address. Figure 12-2 shows a drawing of an Allen-Bradley PLC-5, with 128 I/O. (Power supply is not shown.)

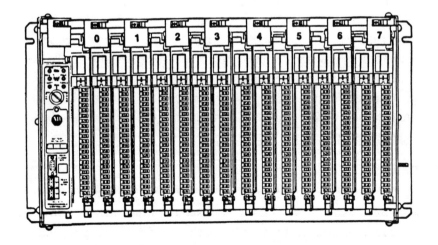

Figure 12-2 PLC-5 with 128 I/O (Courtesy Allen-Bradley Co., Milwaukee WI)

I/O modules are available in many different configurations, and voltages, (AC and DC). Special modules are available to read analog signals and produce analog outputs, provide communication capabilities, interface with motion control systems, etc. The input modules provide isolation from the "real world" control voltages, and give the CPU a continuous indication of the on/off status of each input termination. Inputs sense the presence of voltages at their terminals, and therefore usually have very low current requirements. A PLC-5 I/O rack is shown installed in the lower right compartment of Figure 7-1

Output modules receive commands from the CPU and switch isolated power on and off at the output terminals. Output modules must be capable of switching currents required by the load connected to each terminal, so more attention must be given to current capacity of output modules and their power supply.

Programming devices

Every brand of PLC has its own programming hardware. Sometimes it is a small hand-held device that resembles an oversized calculator with a liquid crystal display (LCD). See Figure 12-3.

Computer-based programmers typically use a special communication board, installed in an industrial terminal or personal computer, with the appropriate software program installed. Computer-based programming allows "off-line" programming, where the programmer develops his logic, stores it on a disk, and then "down-loads" the program to the CPU at his convenience. In fact, it allows more than one programmer to develop different modules of the program. An industrial computer terminal is shown in Figure 12-4.

Programming can be done directly to the CPU if desired. When connected to the CPU the programmer can test the system, and watch the logic operate as each element is intensified in sequence on the CRT when the system is running. Since a PLC can operate without having the programming device attached, one device can be used to service many separate PLC systems. The programmer can edit or change the logic "on-line" in many cases. Trouble shooting is greatly simplified, once you understand the addressing system.

Every I/O point has a corresponding address in the CPU memory. To understand the addressing scheme we need to examine the various number systems that may be encountered in PLC products. Chapter 13 makes a comparison of these systems.

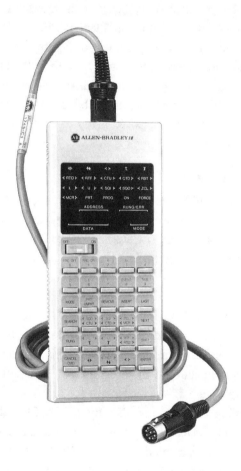

Figure 12-3 Hand-held programmer for small logic controller, SLC 100
(Courtesy Allen-Bradley Co., Milwaukee WI)

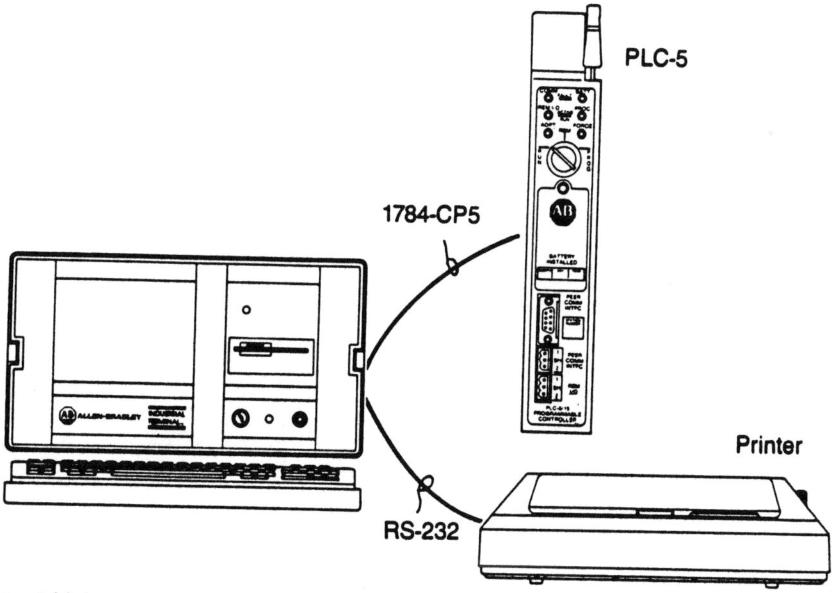

PLC-5

1784-CP5

Printer

RS-232

LC-5 Direct Connect Link
H+ Communication Link)

Figure 12-4   Industrial computer terminal (Courtesy Allen-Bradley Co., Milwaukee WI)

Chapter 13

Number Systems

We are all familiar with our decimal numbering system. It probably originated from the fact that people have 5 digits on each hand, and it was easy to count on our fingers. It has been said that if we had only 4 digits on each hand, we might have had numbering system based on 8 instead of 10. However, when working with computer-type equipment, we find that there are several strange, but more efficient, number systems that allow better utilization of resources. In Chapter 11, we introduced the binary system when discussing methods of coding and decoding information. In this chapter we will define and compare the most common systems.

DECIMAL

The use of decimal numbers is the most common system in the world. Metric measurements are a good example of decimal notation. It is based upon the powers of 10. When written, the numbers are arranged in columns. The rightmost column contains the least significant digit (LSD), having a weight of 10^0, so the numbers are multiplied by 1. The next column has a weight of 10^1, with the numbers multiplied by 10. The third column is 10^2, so all numbers are multiplied by 100, etc. The leftmost column contains the most significant digit (MSD). Following is an example:

	MSD						**LSD**
Power of 10	10^3		10^2		10^1		10^0
Digits	9		7		4		6
Dec. Values	9000	+	700	+	40	+	6 = 9746

The sum of the values is equal to the real decimal number.

BINARY

The binary system has a base of 2. As discussed earlier, a *bit* of logic, or relay has only two conditions, OFF (0) or ON (1). 1 is the largest digit that can be used. The rightmost column has a weight of 2^0, and each digit is multiplied by 1. The next column is weighted at 2^1, and multiplied by 2. The third column has a weight of 2^2, and is multiplied by 4. The fourth column has a weight of 2^3, and is multiplied by 8.

Example:

	MSD						**LSD**
Power of 2	2^3		2^2		2^1		2^0
Digits	1		0		1		1
Dec. Values	8	+	0	+	2	+	1 = 11

Computer-type equipment depends upon the electrical state of discrete bits of electrical information. See Figure 13-1. *A bit is the smallest increment of information that is available.* An electrical signal can have only two (2) states – ON or OFF. Consequently, all other numbering systems must ultimately be derived and encoded from binary information.

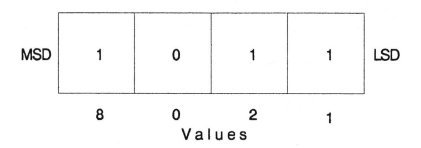

Figure 13-1 Binary bit pattern

OCTAL

Octal numbers are usually used for storing addresses. There is a total of 8 bit patterns that can be derived from 3 bits of information: 0,1,2,3,4,5,6, and 7. If we count from 0 and 7, we use every possible combination available with a block of three (3) bits with <u>nothing left over</u>. In other words, if we count using a base of 8, we will have the most efficient use of our memory for storing addresses. Note that the binary values below, use blocks of three digits.

Octal counting compared with decimal:

Octal	=	Decimal	Binary	Octal	=	Decimal	Binary
0	=	0	000	10	=	8	001 000
1	=	1	001	11	=	9	001 001
2	=	2	010	12	=	10	001 010
3	=	3	011	13	=	11	001 011
4	=	4	100	14	=	12	001 100
5	=	5	101	15	=	13	001 101
6	=	6	110	16	=	14	001 110
7	=	7	111	17	=	15	001 111

The octal system is used extensively in identifying I/O points as well as memory locations. See Figure 13-2.

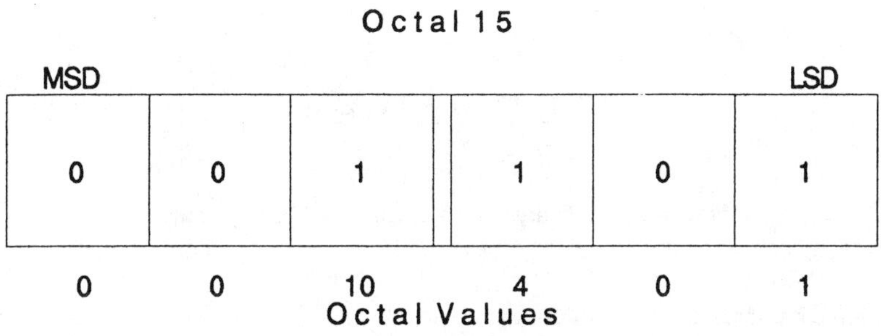

Octal 15

MSD					LSD
0	0	1	1	0	1

| 0 | 0 | 10 | 4 | 0 | 1 |

Octal Values

(Since Octal has no #8 or #9, this is equivalent to a Decimal 13)

Figure 13-2 Octal bit pattern

BINARY CODED DECIMAL (BCD)

Binary coded decimal (BCD) numbers provide read-outs in the familiar decimal format. In this case, four (4) bits of information are required for each digit. There are 16 patterns that can be derived from four bits, but only the first ten patterns are used, giving us 0 through 9. While this is more convenient to read, it does waste some resources.

BCD encoding (See Figure 13-3):

BCD	Binary	BCD	Binary
0	0000	10	0001 0000
1	0001	11	0001 0001
2	0010	12	0001 0010
3	0011	13	0001 0011
4	0100	14	0001 0100
5	0101	15	0001 0101
6	0110	16	0001 0110
7	0111	17	0001 0111
8	1000	18	0001 1000
9	1001	19	0001 1001

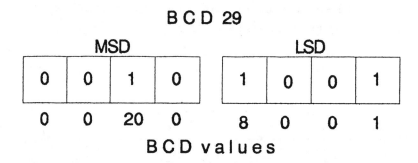

Figure 13-3 Binary Coded Decimal (BCD) bit pattern

HEXADECIMAL

Hexadecimal numbers have a base of 16. This is the total number of binary combinations that one can get from a block of 4 bits of information. In this case single digit numbers are not limited to a maximum of 0 through 9, but continue with letters of the alphabet, utilizing all available resources. A comparison of hexadecimal and decimal counting is shown below.

Hexadecimal count	0	1	2	3	4	5	6	7	8	9	A	B	C	D	E	F
Decimal count		0	1	2	3	4	5	6	7	8	9	10	11	12	14	15

Instead of counting 0 through 9 as in the decimal, each column counts 0 through F, which is the equivalent of 0 through 15 decimal. As each 4-bit column is added, they increase by a power of 16. See following example.

Hexadecimal										Decimal
Powers of 16	16^3		16^2		16^1		16^0			
1111 =	4096	+	256	+	16	+	1	=		4369
FFFF =	61440	+	3840	+	240	+	15	=		65535
F10A =	61440	+	256	+	0	+	10	=		61706

With four columns of hexadecimal numbers, any number from 0 to 65535 can be represented. One *word* of 16 bits can provide 65536 different numbers. See Figure 13-4. These numbers are used to map and address large blocks of memory.

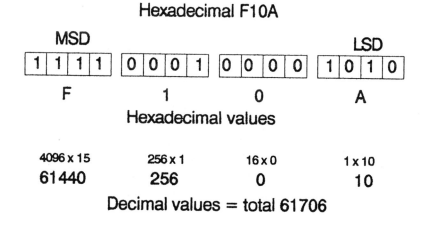

Hexadecimal F10A

Figure 13-4 Hexadecimal bit pattern

These various number systems are a bit strange to deal with, and are sometimes hard to understand, but this gives one a tool for interpreting them when they are encountered. As we discuss memory mapping in the following chapter, one will see how some of them are applied.

Chapter 14

PLC Memory Structure

The memory structures of different PLCs are quite similar in concept, but you will find variations in addressing and the way they are displayed. The structure that we will use for our examples relates to Allen-Bradley PLC-5 equipment. There is no attempt to fully describe this product, as its capabilities go far beyond the examples we will use. Memory structures are divided into the following increments.

BIT

A *bit* is the smallest increment of information that computer-type equipment can work with. As mentioned in earlier discussions, a bit has only two states. When it is energized, the state is true, or 1. When it is not energized, the state is false, or 0.

BYTE

Eight (8) bits make up a *byte*. A byte is the smallest increment of information that can transmit an ASCII character. ASCII is the abbreviation for American Standard Code for Information Interchange, and is used to transmit data between computers and peripheral equipment. Each letter, space, number, and character in this book requires a byte of information so that it can be sent to a printer.

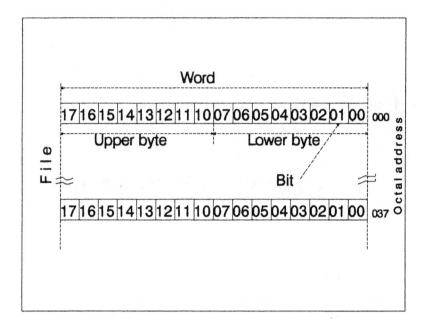

Figure 14-1   Bit, Byte, Word, File relationships

WORD

A *word* is made up of sixteen (16) bits or two (2) bytes. The byte with the lowest address numbers is called a *lower byte*, and the one with the higher addresses is referred to as an *upper byte*. See Figure 14-1.

The sixteen bits may have octal or decimal addressing. In our example, words related to I/O have octal addressing, 0 through 7 and 10 through 17. Other words have decimal addressing, 0 through 15.

FILE

A file contains any number of words. Files are separated by type, according to their functions. Figure 14-2 shows the memory organization of a PLC-5 controller. The files range from 32 to 1000 or more words per file. The Integer and Floating Point files allow the PLC to make complex computations, and display data in Decimal, Binary, Octal, and Hexadecimal formats. Since we will be comparing relay logic with a PLC program, most of our discussion will be directed to Output Image, Input Image, Timer/Counter, and Main (user logic) files.

Memory Organization

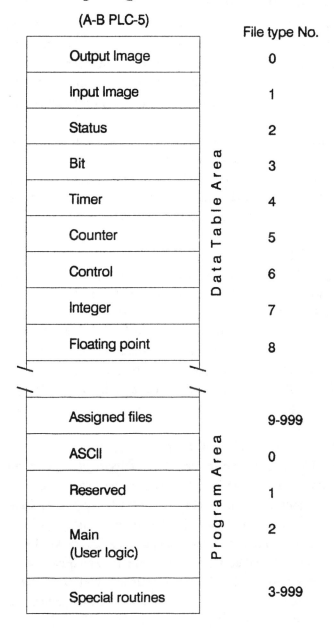

Figure 14-2 Memory organization for Allen-Bradley PLC-5

SYSTEM ADDRESSING

The key to getting comfortable with any PLC is to understand the total addressing system. We have to connect our discrete inputs, push-buttons, limit-switches, etc., to our controller, interface those points with our "electronic ladder diagram" (program), and then bring the results out through another interface to operate our motor starters, solenoids, lights, etc..

Inputs and outputs are wired to interface modules, installed in an *I/O rack*. Each rack has a two-digit address, each slot has its own address, and each terminal point is numbered. With this product, all of these addresses are octal. We combine the addresses to form a number that identifies each input and output.

I/O Addresses

Figure 14-3 shows a very simple line of logic, where a push-button is used to turn on a lamp. The push-button and lamp "hard-wiring" terminates at I/O terminals, and the logic is carried out in software.

We have a push-button, wired to an input module (I), installed in rack 00, slot 0, terminal 04. The address becomes *I:000/04*.

An indicating lamp is wired to an output module (O), installed in rack 00, slot 7, terminal 15. The address becomes *O:007/15*.

Image table addresses

File 0 in the memory is reserved as an *output image table*. (See Figure 14-2). Every word in file 0 has a three-digit octal address beginning with 0:, which can be interpreted as the letter O for *output*. The word addresses start with octal *0:000* to *0:037* (or higher). This number is followed with a slant (/) and the 2-digit bit number.

File 1 is reserved as an *input image table*. Every word in file 1 has a three-digit octal address beginning with 1:, which can be interpreted as the letter I for *input*. The word addresses start with octal *1:000* to *1:037* (or higher). This number is also followed by a slant (/) and the 2-digit bit number.

I/O Rack

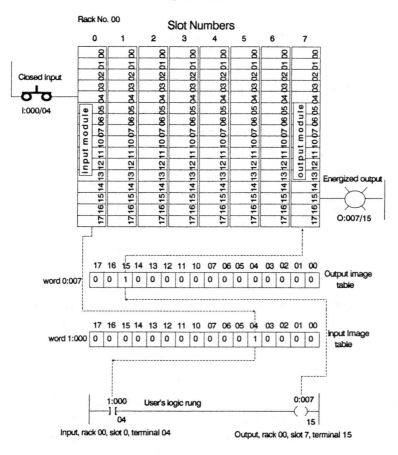

Figure 14-3 Solution of one line of logic

Our input address, I:000/04, becomes memory address 1:000/04, and the output address O:007/15 becomes memory address 0:007/15. In other words, type of module, rack address, and slot position identify the word address in memory. The terminal number identifies the bit number.

If you know the address of the input or output, you can immediately check the status of its bit by calling up the equivalent address on the CRT screen.

SCANNING

When running, the CPU scans the memory continuously from top to bottom, and left to right, checking every input, output, and instruction in sequence. The scan time depends upon the size and complexity of the program, and the number and type of I/O. The scan may be as short as 15 milliseconds or less. 15 milliseconds per scan would produce 67 scans per second. This short time makes the operation appear as instantaneous, but one must consider the scan sequence when handling critically timed operations and sealing circuits. Complex systems may use interlocked multiple CPUs to minimize total scan time.

As the scan reads the input image table, it notes the condition of every input, then scans the logic diagram, updating all references to the inputs. After the logic is updated, the scanner resets the output image table, to activate the required outputs.

In Figure 14-3, we have shown how one line of logic would perform. When the input at *I:000/04* is energized, it immediately sets input image table bit *1:000/04* true (on). The scanner senses this change of state, and makes the element *1:000/04* true in our logic diagram. Bit *0:007/15* is turned on by the logic. The scanner sets *0:007/15* true in the output image table, and then updates the output *0:007/15* to turn the lamp on. Figure 14-4 shows some optional I/O arrangements and addressing.

Some manufacturers use decimal addresses, preceded by an X for inputs and Y for output. Some older systems are based on 8-bit words, rather than 16. There are a number of proprietary programmable controllers applied to special industries, such as elevator controls, or energy management, that may not follow the expected pattern, but they will use either 8 or 16 bit word structures. It is very **important** to identify the

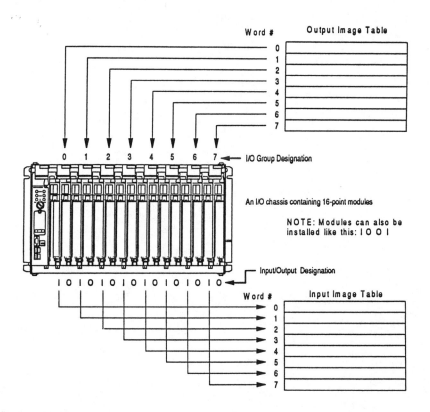

Figure 14-4 I/O addressing scheme

addressing system before you attempt to work on any system that uses a programmable controller, because one must know the purpose of each I/O bit before manipulating them in memory.

To illustrate the logical power of a PLC, the following chapter summarizes a typical instruction set. It will be obvious that there is the capability of developing very complex programs far beyond the scope of our discussions.

Chapter 15

PLC Instruction Set

TYPES OF INSTRUCTIONS

There are several types of instructions that are available in a PLC. The instruction set will vary depending upon the brand and type of processor that is used. A PLC-5 has nine (9) types of instructions.

1. *Bit level* – Bit level instructions include *examine on, examine off, output energize, output latch, output unlatch* and *one shot*. These are the most common instructions used for simple ladder diagrams and are treated as <u>relay</u> logic. Combinations of *examine on* and *examine off* instructions with branching, appear as NO and NC contacts respectively. The *output* instructions appear as coils in the diagram.

2. *Word level* – There are over twenty (20) word level instructions that use sixteen (16) bit words for calculations, comparisons, conversions, and masking. These instructions appear as blocks, identifying the word addresses, and the desired function, activated by one or more *bit level* instructions.

3. *File level* – These instructions allow the use of multiple word files to perform *file arithmetic and logic, file search and compare, sequencer output, sequencer input,* and *sequencer load functions.*

4. *Timers and Counters* – Timers and counters include *timer on-delay, timer off-delay, retentive timer, up-counter, down counter,* and *reset.* These instructions reserve three (3) sixteen bit words (total 48 bits) for their respective functions in a special file. They also appear as block instructions that display time-base, preset values and accumulated values as the instruction operates. These instructions are controlled by *bit level* instructions.

5. *Program control* – There are ten (10) program control instructions used for such functions as sub-routines, label, jump, immediate inputs and outputs and related actions. These appear as block instructions in the ladder diagram.

6. *Communications* – Three (3) instructions allow *block transfer* reads and writes, and *message* handling. These allow the exchange of data between remote processors and I/O racks.

7. *Shifting* – Five (5) instructions allow manipulation of bits within words. These include the ability to perform FIFO (first-in, first-out) functions.

8. *Diagnostic* – Three (3) instructions include *file bit compare, diagnostic detect,* and *data transitional functions.*

9. *Miscellaneous* – PID which provides a *proportional, integral, derivative* control that can be tuned to match process requirements.

This is a very powerful instruction set that requires a lot of training to fully utilize its total capabilities. Most simple operations utilize the *bit level* and *timer/counter* instructions, and when printed out, appear very similar to a relay ladder diagram.

Most electricians, who are familiar with relay circuits, can usually trouble-shoot the simpler PLC systems without formal programming experience. However, they must learn the operating techniques required to manipulate the programming and monitoring system without causing faults. As mentioned earlier, it is most important to understand the addressing system so one can recognize the function of each symbol shown on the PLC programming screen when trouble-shooting. Following are definitions of many of the available instructions.

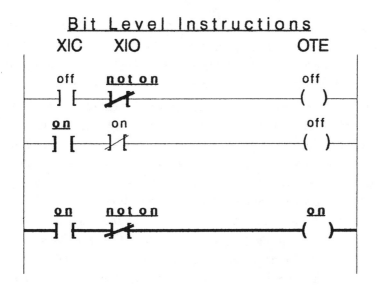

Figure 15-1 Bit-level instruction logic

BIT LEVEL INSTRUCTIONS

Let's look first at bit level instructions that are the most common instructions found in the simpler PLC installations. These instructions follow the general pattern of the ladder diagrams related to relay logic. See Figure 15-1.

Examine on is often abbreviated XIC, appears as a NO contact. If the element is not energized, the contact appears on the CRT screen with normal intensity. When <u>energized</u>, the contact will appear <u>intensified</u>, indicating that we have a conducting element in our ladder diagram. The element can be an input, or be derived from an output, or relay function within the program. The fact that this is an *examine* instruction means that we can look at the status of any input or output element as often as necessary, and add it to a new rung. If this were a relay ladder diagram, we would have to add another isolated contact each time we needed to include the status of any relay.

Examine off, abbreviated *XIO,* appears as a NC contact. If the element is <u>not energized</u>, the contact will appear as a conducting element,

intensified on the screen. If it changes to an energized state, the element
will revert to lower intensity, appearing as a non-conducting element in
our ladder diagram. As above, we can examine any other input or output as often as necessary to determine that it is not energized.

Output energize, (OTE) will be the last element on the right side of
our ladder diagram, and symbolize a coil. When all necessary conditions
in the rung are "true" (conducting), the output will appear intensified.
This will turn on any load that is connected to its terminal in the I/O
rack. We can add as many *XIC* and *XIO* instructions as we need in our
program, addressed from this output, to show its status.

Output addresses that are not assigned to an I/O rack can be used as
internal relays within the program. This is useful in cases where a complex set of conditions are required in many rungs of logic. One can
build one complex rung, terminating with a relay address, then merely
examine that relay's status with a single "contact" in each rung that
requires its logic, simplifying the program, and saving memory.

Output latch (OTL) and *Output unlatch (OTU)* instructions are
always used in pairs with the same address. When conditions are true in
an *OTL* rung, the output will turn on, and remain on after the conditions
turn false (off). A separate rung must be used to activate an *OTU*
(unlatch), with the same address, to reset the original *OTL* to an
unlatched condition.

One shot (ONS) outputs only stay on for one scan after the rung is
true. This provides a pulse to control a following rung. The rung must
go false before another pulse can be repeated. This is a useful instruction, often used to provide a "triggering" signal when an object first
covers a limit switch, without holding a maintained signal that might
interfere with following logic. In other words you can have a single
short pulse from an input that may be held energized for a sustained
time interval.

WORD LEVEL INSTRUCTIONS

Following are brief descriptions of word level instructions used
principally to work with numbers and shift registers. Analog input and
output values are usually stored and manipulated as words.

Compute (CPT) instruction performs copy, arithmetic, logical and conversion operations. It is an output instruction that performs operations you define in an expression, and writes the result into a destination address. It can copy data from one address to another and convert the source data to the data type specified in the new address (e.g. convert octal numbers to hexadecimal, etc.).

Add (ADD) instruction adds one value to another value, and places the result in a destination address.

Subtract (SUB) instruction subtracts one value from another value, and places the result in a destination address.

Multiply (MUL) instruction multiplies one value by another value, and places the result in a destination address.

Divide (DIV) instruction divides one value by another value, and places the result in a destination address.

Square Root (SQR) instruction takes the square root of a value, and place the result in a destination address.

Negate (NEG) instruction is used to change the sign (+ or -) of a value.

BCD to Integer (FRD) instruction converts a BCD value to an integer value.

Integer to BCD (TOD) instruction converts an integer value to a BCD value.

Move (MOV) instruction copies a value from the source address to a destination every scan. Often used to pick up a continuously changing input address and place it into a continuous computation address.

Move with mask (MVM) instruction is similar to MOV except the move passes through a mask to extract bit data from an element that may contain bit and word data.

Clear (CLR) instruction is used to reset all bits in a word to zero.

Bit inversion (NOT) instruction performs a NOT operation using the bits in a source address. It inverts the status of bits, i.e., a (1) becomes a (0), and a (0) becomes a (1). Figure 15-2 illustrates "truth tables" used to define several of these related instructions.

Logic Truth Tables

AND

Input		output
x	y	z
0	0	0
0	1	0
1	0	0
1	1	1

A function that produces a ONE (1) state if and only if each input is in ONE (1) state.

OR

input		output
x	y	z
0	0	0
0	1	1
1	0	1
1	1	1

A function that produces a ONE (1) state if and only if one or more inputs are a ONE (1) state.

XOR

input		output
x	y	z
0	0	0
0	1	1
1	0	1
1	1	0

A function that produces a ONE (1) state If and only if one, but not both inputs are a ONE (1) state.

NOT

input	output
x	z
0	1
1	0

A function that inverts an input to opposite state output. A ZERO (0) state input produces a ONE (1) state output. A ONE (1) state input produces a ZERO (0) state output.

Figure 15-2 Logic truth tables

Bit-wise and (AND) instruction performs an AND operation using the bits in two source addresses. The result is placed in a destination address. See Figure 15-2.

Bit-wise or (OR) instruction performs an OR operation using the bits in two source addresses. The result is placed in a destination address. See Figure 15-2.

Bit-wise exclusive or (XOR) instruction performs an exclusive OR operation using bits in two source addresses. The result is placed in a destination address. A (1) is placed in the destination address if corresponding bits in source addresses are not equal. See Figure 15-2.

Compare (CMP) is an input instruction that compares values and performs logical comparisons. When comparison is true, the rung becomes true.

Equal to (EQU) is an input instruction that tests whether two values are equal, and makes the rung true when the values are equal.

Not equal to (NEQ) is an input instruction that tests whether two values are not equal, and makes the rung true when the values are not equal.

Less than (LES) is an input instruction that tests whether one value is less than a second value, and makes the rung true when the conditions are true.

Less than or equal to (LEQ) is an input instruction that tests whether one value is equal to or less than a second value, and makes the rung true when the conditions are true.

Greater than (GRT) is an input instruction that tests whether one value is greater than a second value, and makes the rung true when the conditions are true.

Greater than or equal to (GEQ) is an input instruction that tests whether one value is equal to, or greater than a second value, and makes the rung true when the conditions are true.

Circular limit test (LIM) is an input instruction that tests whether one value is within the limits of two other values. Rung is set true when condition is true.

Masked equal to (MEQ) is an input instruction that passes two values through a mask, and test whether they are equal. Rung is set true when condition is true.

FILE LEVEL INSTRUCTIONS

File arithmetic and logic (FAL) is an output instruction that performs copy, arithmetic, logic, and function operations on the data stored in files. It performs the same operations as CPT instruction, except FAL works with multiple words, rather than single words.

File search and compare (FSC) is an output instruction that performs search and compare operations on multiple words. Performs same operations as CMP which is limited to single words.

SEQUENCERS

Sequencer instructions are used to control automatic assembly machines that have a consistent and repeatable operation using 16 bit data.

Sequencer output (SQO) is an output instruction that steps through a sequencer file of 16-bit output words whose bits have been set to control various output devices. When the rung goes from false to true, the instruction increments to the next (step) word in the sequencer file.

Sequencer input (SQI) is an input instruction used to monitor machine operating conditions for diagnostic purposes by comparing 16-bit image data, through a mask, with data in a reference file.

Sequencer load (SQL) is an output instruction used to capture reference conditions by manually stepping the machine through its operating sequences and loading I/O or storage data into destination files. (A machine "teaching" tool).

TIMER AND COUNTER INSTRUCTIONS

Timers and counters are shown in the position of outputs in the PLC print-outs. As mentioned earlier, they appear as block instructions, and are controlled by *XIO, XIC,* or other contacts in their respective rung of logic. The *done (DN)* bits of the timers and counters are energized when timing or counting values match the preset values, and are used as contacts in other rungs as required – as often as necessary.

Timers also have a *Timer timing (TT)* bit that is energized whenever timing is in progress. *Energized* bit *(EN)* is true (on) whenever the rung is true. See Figure 15-3.

Timer on-delay (TON) will have a selectable time base (e.g. 0.01, 0.1, 1.0 seconds), a selectable preset showing the number of increments of the time base to be used for the time delay, and the accumulated value while the timer is running. When the accumulated value equals the preset, the *DN* bit will turn on.

Timer off-delay (TOF) will have the same functions, but will turn on when its rung is true, and hold its *DN* bit energized for the preset time after its rung turns false (off).

Example of TON ladder diagram

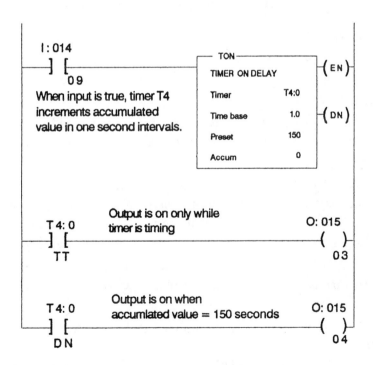

Figure 15-3 On-delay timer ladder diagram

Retentive timer (RTO) operates similar to a *TON,* but accumulates time whenever its rung is true, holds that value when the rung becomes false, and then resumes timing where it left off each time the rung is true. When the accumulated value equals the preset, the *DN* bit is energized and held on. A separate rung must be provided to control a *Reset* instruction *(RES)* having the same address as the *RTO.* When the *RES* instruction is true, it resets the *RTO* accumulated value to zero.

Count up (CTU) and *Count down (CTD)* instructions count up or down one increment each time their respective rung is energized. A *DN* bit is set when the accumulated value is equal to the preset. *CTU* and *CTD* instructions can be used in pairs, having the same address. A *RES* instruction with the same address is used to reset the accumulated value to zero, similar to the *RTO* instruction. An *Overflow* bit *(OV)* is turned on if the count exceeds the preset of the counter. All these bits can be used where needed, anywhere in the program, and as often as necessary.

PROGRAM CONTROL

Immediate input (IIN) is an output instruction used to update a word of input image bits before the next normal update, by interrupting program scan momentarily.

Immediate output (IOT) is an output instruction used to update an I/O group of outputs before the next normal update, by interrupting program scan momentarily.

Master control reset (MCR) instructions are used in pairs to create program zones that turn off all non-retentive outputs in the zone. Allows one to enable or inhibit segments of the program as might be needed in recipe applications.

Jump (JMP) is an output instruction that allows one to skip portions of the ladder logic. It is used with LBL which designates the destination of the jump.

Label (LBL) is an input instruction that identifies the destination of a JMP instruction.

Jump to subroutine (JSR)
Subroutine (SBR)
Return from subroutine (RET)

JSR, SBR, and RET instructions direct the processor to go to a separate subroutine file within the ladder processor, scan that subroutine file once, and return to the point of departure.

Temporary end (TND) is an output instruction that returns the scan to the beginning of the program. Often inserted progressively through the program for debugging and trouble-shooting a program.

Always false instruction (AFI) is an input instruction that can be temporarily used to disable a rung during testing and trouble-shooting.

COMMUNICATIONS INSTRUCTIONS

Block transfer read (BTR) is an output instruction that can transfer up to 64 words at a time from an I/O chassis or supervisory processor.

Block transfer write (BTW) is an output instruction that can transfer up to 64 words at a time to an I/O chassis or supervisory processor.

Message (MSG) is an output instruction used to send "packets" of up to 1000 elements of data between processors on a data highway network.

SHIFTING INSTRUCTIONS

Bit shift left (BSL) used to move bits one address left.

Bit shift right (BSR) used to move bits one address right.

Bit field distributor (BTD) used to move bits within a word.

Fifo load (FFL) used to load a first in, first out file, storing words in a "stack." Used with *FFU*.

Fifo unload (FFU) used to provide first in, first out operation, removing words from an *FFL* "stack."

Lifo load (LFO) used to load a last in, first out file, storing words in a "stack." Used with *LFU*.

Lifo unload (LFU) used to provide last in, first out operation, removing words from an *LFO* "stack."

The above instructions are output instructions used to move bits in various modes for shift register operations. Can be used to track product as it moves down a processing line.

DIAGNOSTIC INSTRUCTIONS

File bit compare (FBC) is used to compare I/O data against a known, good reference, and record mismatches.

Diagnostic detect (DDT) is used to compare I/O data against a known, good reference, record mismatches, and update the reference file to match the source file.

Data transitional (DTR) is used to pass source data through a mask and compare to reference data, then write the source word into the reference address of the next comparison.

MISCELLANEOUS INSTRUCTIONS

Proportional, integral, derivative (PID) is a complex instruction used for process control for such quantities as pressure, temperature, flow rate, and fluid levels. Inputs are brought in through A/D converters, and processed as integer numbers. After comparative calculations are made, the integer numbers are sent out as outputs to D/A converters for corrective control.

SUMMARY

The instruction set listed in this chapter is not all inclusive nor have we tried to document all the ways these instructions can be applied. Bit level, relay instructions and timer/counter instructions are the most common elements that will be found in single machine operations. As production lines become more complex, with inter-dependent operations between machines, you will find more and more of the complex instructions used. This list will help one identify the type of logic being processed, if some of the more complex instructions are encountered.

Since we are comparing simple PLC programs to ladder logic, we will limit our applications to the more common instruction sets described above. You will find similar instructions in most brands of PLCs. The configuration of timers and counters may appear differently, but the same functions will usually be present.

One can visualize a simple PLC program as a relay schematic whose relays have a limitless number of contacts. The logic, however, examines the state of each relay, not actual hardware, and we just add another contact to the schematic wherever required.

Chapter 16

How Do PLCs Compare to Relay Logic?

In early chapters, we discussed ways we apply logic to design and trouble-shoot electrical ladder diagrams. We also discussed the principal characteristics of programmable logic controllers and the available instructions sets. In this chapter, we will illustrate how this logic can be compared with a PLC program.

In Chapter 3, the schematic of a motor control circuit with a run/jog option was illustrated in Figure 3-1. We will use this simple circuit to illustrate a conversion to PLC control with some enhancements.

PLC installations require "hard-wired" components, terminating at input and output terminals, where they interface with the software program. This means that we have to provide documentation for both the hard-wiring and the software logic.

The most difficult hurdle we face in trouble-shooting a PLC installation is interpreting the designer's interface addressing. A high percentage of industry's documentation requires switching back and forth between multiple drawings, print-outs, and instruction books to identify which component (or function) is represented at each input. On the other hand, we have the same problem identifying what happens as each output is energized.

Some vendors provide drawings of the hard-wired components, using only their abbreviated identification, terminating at input addresses with no explanation of the functions. The software is provided as a computer print-out which often has no annotation to identify function of each rung. This requires one to look up a table that may be buried in the instruction book, or on an obscure drawing to learn what each abbreviation means. It is so cumbersome that the logic might as well be considered proprietary, as only the manufacturer can decipher it with any expedience.

Annotation and element indexing are available from most PLC manufacturers, and should be supplied to the customer. If a consistent pattern of properly labeled hard-wired components and program elements is followed, a comparison of the schematic with a separate annotated program print-out can be considered an acceptable documentation for trouble-shooting.

Figure 16-1 illustrates a documentation system that is often incorporated in industrial machine controls. The inputs are shown as hard-wired components connected to the high side of the control circuit, and terminating at PLC input addresses. Each input appears as close as possible to the related ladder rung in the program running down the center of the drawing. Each rung is annotated to identify its purpose.

Outputs are shown, picking up their power from their respective output addresses, and terminating at the low side of the control circuit.

Rungs can be numbered similar to relay logic ladder diagrams, and the locations where each input and output is examined are noted at each terminal point on the drawing.

ANALYZE THE SAMPLE PROGRAM

Some unique conditions that must be considered when using a PLC are illustrated in Figure 16-1. As we trace through the logic, one will note that we are "examining" the status of each input and output terminal, not looking at the discrete hardware component.

On the left of our drawing, we have shown a NC *Stop* push-button, a NO *Start push-button,* a NO *Run/Jog selector* (run position), and a NC *motor overload contact.* Each of these connect to an input address, corresponding to their I/O rack terminal location.

Rung #1 starts with an apparently NO *XIC* contact element, addressed to the NC Stop push-button, followed by another XIC

MOTOR CONTACTOR
CIRCUIT

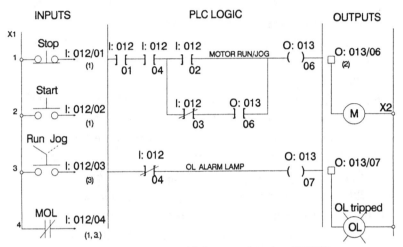

Note that the NC inputs for the STOP
pushbutton and the Motor OverLoad
are programmed as XIC contacts because
the inputs must be held closed when running
and stop the motor when de-energized.

Figure 16-1 Motor contactor circuit, using PLC logic

addressed to the NC *motor overload contact.* One might ask, "Why are we showing NC inputs as NO elements in our program?" You will recall that an *XIC* element becomes true (conducting) when its input address is true. If the NC *Stop* push-button does not change state, the input *I:012/01* will be energized and make the *XIC* true in the ladder diagram. If the push-button is opened, the address becomes false, and de-energizes the rung to shut down the motor.

The same logic applies to the *motor overload contact* input *I:012/04.* If an overload occurs, this address will be de-energized, and its *XIC* will de-energize the rung, and shut down the motor.

With the first two elements true, the next bit of logic is examining the *Start push-button* input at *I:012/02.* If we press this NO push-button, we make all of the elements true that are required to energize the output *O:013/06.* This will activate the *XIC* element with the same output address in the lower branch of this rung, and if *I:012/03* is not true, the output will remain energized when the *Start push-button* is released.

If the *Run/Jog selector switch* is turned on, to select the jog function, *I:012/03 XIO* will stop conducting, and the output can only be true when the *Start push-button* is held in a true condition. It cannot seal in.

If we did not bring the *motor overload* contact in as an input, the motor contactor *M* could drop out on overload, even though the output logic showed it to be energized. We added another rung at line 3 that will turn on an output, addressed to activate a warning light at *O:013/07,* when input *I:012/04* goes false from a tripped overload.

With this rung, we can tell why the motor contactor shut-down. If an overload occurs, *O:013/06* will turn off, and *O:013/07* will be turned on.

Many levels of "branching" can be included in a single rung of logic. Very complex multi-path input structures can be assembled, controlling a single output, or sometimes multiple outputs. In lengthy programs, an expert programmer may save programming memory, and optimize scan time by using this capability. However, this complexity does not lend itself to easy trouble-shooting by someone who is unfamiliar with the program logic. If memory is adequate, and scan time is not critical, it is adviseable to break simple programs into more rungs of simple logic that can be adequately annotated, and understood by more people.

Large applications, requiring multiple interlocked PLCs, require personnel who have PLC communicating skills. It is unlikely that a maintenance person who has not had extensive training would be expected to take the responsibility those complex systems. Most electricians can pick up the needed knowledge to work on systems that emulate relay logic, and those are the people this material is directed to.

Chapter 17

Convert Relay Logic to PLC

In Chapter 6, we developed the relay schematic diagram for the conveyor control system. We will use the same application to illustrate how it can be represented in a PLC system. Compare Figure 6-4 (or Figure 9-1) with Figures 17-1, 17-2, and 17-3 as the discussion continues.

POWER CONSIDERATIONS

In an earlier chapter, we mentioned the fact that most inputs provide a "voltage present" signal to the input modules. Current is really not a problem in most instances, but voltage quality is important. The CPU requires a power source that is as free as possible of line disturbances. Since there are minimum disturbances created by input signals, it is common practice to use a common power source for the CPU and inputs.

However, outputs must switch loads with varying currents. Often a large contactor or solenoid will not only draw a high in-rush current, but also create a high-voltage inductive voltage spike when de-energized. If the CPU is subjected to these disturbances, its low-level logic components may be damaged, followed by erratic operations or premature failure. On large installations, a separate load transformer is recommended for the outputs, isolating the inputs and CPU from load-created line disturbances.

When working with solid state devices, we must remember that there are "leakage currents" present even during apparent OFF conditions. If an element fails, the leakage currents may increase to a point where it may fault in the ON mode. This failure may turn ON other logic, causing dangerous, unpredictable operations. For this reason, a "hard-wired" *emergency stop* circuit should always be installed to shut-down the output power supply in the event of an electronic fault.

Figure 17-1 illustrates one way the above requirements can be shown. Comparing this schematic to Figure 6-4, you will see that we have a control transformer *CX,* and a separate load transformer *LX.* Each transformer has its own output control switch. The control circuit has "pull-on, push-off" push-button controlling the system's only relay *ES,* whose contacts control the power to the input circuits and output modules. This allows de-energizing all inputs and outputs, while main-taining power to the CPU for check-out and trouble-shooting.

LOGIC CONSIDERATIONS

The input and output modules have been shown as simple blocks to show power wiring. The *input* terminal addresses shown on the left side of the schematic represent terminals on the input modules. The outputs on the right side of the schematic receive their power from *output* termi-nal addresses on the output modules.

The programmed logic is shown in the center of the drawing, and represents the lines of logic that you observe on your CRT screen. An effort is made to show inputs and outputs as close as possible to their first occurrence in the logic diagram. This arrangement helps a trouble-shooter, not familiar with the circuit, to quickly see an obvious pattern of logic.

INDEXING

All hard-wired circuits have been assigned wire numbers. Each line of logic is indexed on the left margin. Inputs and outputs are indexed to show the lines of logic where their elements are used. Each line of logic is annotated to identify its function, either on its line, or adjacent to it.

PROGRAMMING CONSIDERATIONS

On line 13, the three overload contacts are connected in series and connected to input address *1:000/00.* When all contacts are in their nor-

PLC CONVEYOR CONTROL

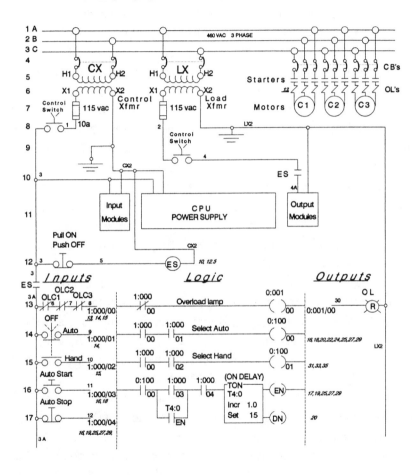

Figure 17-1 PLC conveyor control

mally closed condition input *1:000/00* is true (ON). The first line of logic examines this address for a false (OFF) condition which turns on the output *0:001/00* to energize the OL (overload) indicating lamp. Since the input is normally energized (true), the lamp will not be lit until the input circuit is broken and voltage is not present on wire #8.

Indexing shows that *1:000/00* is also examined on lines 14, and 15, for a true (ON) condition. These three lines must see a true *1:000/00*, indicating no overload condition, before any other logic can proceed. We can examine any address as often as necessary in any line of logic.

On line 14, address *1:000/01* is true when our selector switch is set on *Auto*. If *1:000/00* and *1:000/01* are both true, output *0:100/00* will be true. This output is being used as an internal relay to store this information. Unused output addresses can be used for internal relays. Note that actual inputs and outputs are addressed to rack *00*. The relay output in this rung is addressed to rack *10* which is not being used for real outputs, but is available in the output image table. Note that this "relay" is examined in eight (8) other rungs of logic.

Line 15 has the same logic to store the information that *Hand* has been selected. It is examined in three (3) other rungs of logic.

Line 16 is the first rung that is initiated in the *Auto* mode and starts the on-delay timer controlling the start of conveyor 3. It requires that *Auto* is selected, the *Auto-start* PB pressed, and *Auto-stop* true, to start the timer *T4:0*. The "energized" bit, *T4:0/EN*, is used to seal this rung ON until *Auto-stop* is de-energized. If you look ahead to Figure 17-2, you will see that *T4:0/EN* is also used as a holding contact for the warning circuit on line 19.

Here is an instance where the rung sequence is important. Bits *0:100/01*, *1:000/03*, and *1:000/04* are used in rungs 16 and 18 as start/stop control for both rungs. If the "warning" rung were programmed in a sequence ahead of the *T4:0* timer, that rung would not seal in, because the scan would leave the warning rung before the timer is activated to supply the sealing contact *T4:0/EN*.

Continuing with Figure 17-2, we find the rest of the logic required to provide automatic starting and stopping sequence. Bit *0:100/00* appears in every rung, identifying them as automatic rungs. Off-delay timers *T4:3*, *T4:4* and *T4:5* have identical logic including a *T4:0/EN* bit

PLC CONVEYOR CONTROL
(CONTINUED)

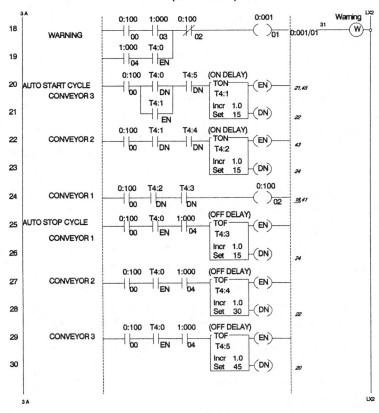

Figure 17-2 PLC conveyor control, continued

and a *1:000/04 auto/stop* bit that keeps them energized during the entire *Auto-run* operating time. The *T4:0/DN* bit in rung 20 starts *T4:1* timer after 15 seconds, and *T4:1/EN* bit turns on *C3* in rung 45, starting conveyor 3. (See Figure 17-3). A *T4:1/EN* bit provides a sealing path parallel to *T4:0*, keeping it energized when *T4:0* is cancelled during *Auto/stop* cycle.

After another 15 seconds, *T4:1/DN* energizes *T4:2* whose *EN* bit turns on *C2* in rung 43, starting conveyor 2.

After another 15 seconds, *T4:2/DN* energizes (relay) *0:100/02* to turn on *C1* in rung 41, starting conveyor 1. *0:100/02*, NC element in rung 18 goes false and shuts down the warning signal.

Pressing the *Auto-stop* PB makes *1:000/04* false, de-energizing rungs 16, 19, 25, 27, and 29. The off-delay timers are de-energized, but *C1*, *C2*, and *C3* hold in. As each timer times out, conveyor 1 drops out after 15 seconds, conveyor 2 drops out after 30 seconds, and conveyor 3 drops out after 45 seconds.

Rungs 31, 33, and 35, are constructed similar to Figure 6-3, with a manual mode bit *0:100/01* leading each rung. Jog functions are controlled by bits related to *1:000/13* input from the *Run/Jog* selector switch.

Interlock contacts for *C1*, *C2*, *C3* have been brought in as inputs. When the respective contactor is energized and pulled in, these inputs will turn on the respective indicating lamp outputs to "prove" contactor action.

Each output, controlling *C1*, *C2*, *C3*, has parallel *automatic* and *manual* control bits derived from the preceding logic, so either system can control the conveyors.

COMPARISONS

If you compare the PLC logic with the relay logic shown in Figure 6-4, you can trace patterns that are almost identical. In many cases, the Logic will be provided as a separate print-out. The inputs and outputs are often annotated where used on the print-out to identify their functions. Most software will also print-out a cross-reference table identifying every I/O and internal relay address, and the rungs where they will be found.

In this case there should be a separate wiring diagram showing the schematic layout of hard-wired components, terminating at respective I/O

PLC CONVEYOR CONTROL
(CONTINUED)

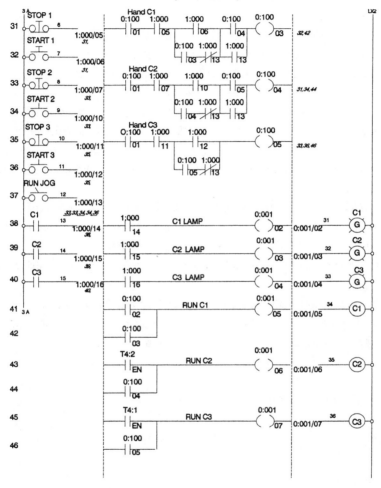

Figure 17-3 PLC conveyor control, continued

addresses. Experience has proven that there are varying degrees of clarity in these interface drawings, depending upon the draftsman's style.

There should always be a chart listing all components, their function, and location in the documentation.

There are a lot of options for the hard-wire layout. As an example, we might have eliminated two (2) inputs if we had used wire #9 from the selector switch as the source of power for *Auto-start* and *Auto-stop* in Figure 17-1, and wire #10 as the source for all "hand" functions in Figure 17-3. Internal relays *0:100/00* and *0:100/01* would not be used, but we would have had to provide additional logic to latch-in *Auto* and *Hand* operations in the rungs that follow. The use of the selector switch inputs in our example, simplifies the programming requirements, because the modes of operation are effectively latched-in at the input terminals.

An experienced programmer could also use short-cuts to pick multiple time intervals from a single timers, by monitoring the "running" bits in the "accumulator" word of the respective timer. While this might save some resources, it would make it more difficult for an "outsider" to decipher his logic.

These examples of programming options emphasize the diversity of logic that one might find for any given application. Patterns of logic are only limited by the imagination of the circuit designer.

Chapter 18

PLC Start-up and Trouble-shooting

Start-up or commissioning of a PLC installation is carried out much like our previous discussion in Chapter 9. The same precautions must be taken to prevent accidental start-up of equipment in the event of a fault. One should lock-out power to motor contactors, and other power devices while testing the logic in a systematic sequence.

The principal difference lies in the use of the Programmer CRT to verify operating tests. Point to point voltage tests can be made with a multi-meter to verify proper hard-wire connections through wires #1 through #4A, Figure 17-1. The PLC can be tested per manufacturer's recommendations for proper set-up, and connections to the programming terminal. The first tests will be made with the output control switch in the Off position so that all outputs are disabled. Consider Figures 17-1 through 17-3 in Chapter 17 as one continuous schematic and program as we continue.

With the PLC in *I/O monitor and control* mode, we bring up each rung of the program in sequence, and operate each input device. This mode may vary from one processor to the next so you must determine how to bring up this screen on your CRT. The indicating lamp should turn ON at each respective input terminal on the I/O rack when the its input is true.

The first rung on line 13 should show no intensified elements if the input from wire #8 at *I:000/00* is energized. If wire #8 is momentarily disconnected, input element *I:000/00* and output element *0:001/00* should intensify, indicating a break in the motor overload circuit. The *I:000/00* elements on lines 14 and 15 should be intensified at all times, unless wire #8 is disconnected.

If we turn our selector switch to *Auto, 1:000/01* should intensify on line 14, and since all elements are true, relay element *0:100/00* should be energized. All automatic rungs from line 16 through 29 should show an intensified element *0:100/00* when *Auto* is selected.

In like manner, if we select *Hand, 1:000/02* should intensify on line 15, and if *1:000/00* is true, relay element *0:100/01* should intensify. All manual references to this element should also be intensified where used on lines 31 through 35 when *Hand* is selected.

Pressing the *Auto Start* push-button when *0:100/00* is energized should turn on the rung at line 16, to start the timer *T4:0,* and energize the warning output *0:001/01* on line 18. As *T4:0* times out, the other on-delay timers will cycle in sequence. As each element becomes true, it will intensify on the screen, and turn on the output for each rung as its required elements become true. The outputs on lines 41 to 45 should come on in their timed sequence. The outputs at lines 38 to 40 will not turn on until output modules are enabled, allowing the contactors to actually pull-in.

The *Auto-stop* cycle can be monitored in the same way. Then test the manual mode inputs.

Most PLC systems also give you the capability to force inputs or outputs on or off to check operations without going to the operator consoles. If someone is monitoring the contactors, with output power on, you can determine if the proper device is connected to each address. The big advantage is that you can prove most of your installation quickly, without running around to watch relays cycle. Since there are fewer hard-wired devices, there are not so many opportunities to have faulty wiring. A copy of the program should be saved on tape or disc, so that it can be quickly restored if a CPU must be replaced.

HISTOGRAMS

Most modern systems give you the capability of developing and printing histograms. In this case, you select any bit address, and literally watch and record the true and false conditions in real time. This is particularly useful if you find intermittent false operations, and have to locate the "culprit" that is causing the problem. It may be a case of an intermittent limit-switch, that is either failing to operate every time, or possibly it is bouncing and providing extra unwanted pulses. It can also help you locate programming errors where a rung is not in its proper sequence so that it misses a sealing or resetting function. This is most likely to happen if more than one programmer is involved in creating the logic.

One illustration of the importance of logic sequence is a replication of the *shift-register* circuits illustrated in Chapter 11. In the relay logic, zone B is used to reset zone A. There is no scanning sequence, so it does not matter what order the ladder rungs are drawn. However, to accomplish this with a programmable controller, all zones have to be programmed in reverse order because zone B must be scanned first to see if it is reset before zone A can send it a new code. This pattern is required for the entire code and decode zone structure, otherwise a complete scan would have to be completed before a reset could be made. In the time for one scan, the code might be overwritten in a following rung before the scan returned to reset the zone. This could totally confuse the logic.

One should think the logic through, one element at a time in "slow motion", always remembering the scanning sequence. A "contact" cannot change state until its line of logic tells it to. If its reference is needed "up-stream" during a single scan, its rung will have to be moved ahead of the required action.

While we have not tried to teach one how to program, it is important to realized that if you modify the program to insert a new rung of logic, its sequence in the program must be considered. You can't arbitrarily "tack it on" at the end of the program. Many program flaws can be attributed to improper scan sequences. It is interesting to note that several manufacturers call their CPU a "Sequencer".

It is obvious that the materials covered in the past seven (7) chapters will not make the reader an experienced PLC programmer. If one can grasp the concept of how the PLC relates to the relay ladder diagram, it should help him cope with almost any programmable controller that might be thrust upon him. Addressing and scan sequence are probably the two most importance concepts that one should understand. Programming techniques can be learned through exposure, but we would certainly advise the technician to seek, as a minimum, a first level factory or distributor-sponsored school where he will be taught how to work with the hardware, as well as the instruction set.

A well-planned PLC installation can save thousands of dollars in materials and installation costs. Maintenance is easier than fighting banks of electro-mechanical relays, and reliability is excellent. It also gives a plant a communication link to its processes that can provide statistical and production information that would not be available with relays. However, the investment in a PLC installation could be in jeopardy if the plant does not also invest in basic training for their maintenance personnel so they can maintain the system intelligently.

Chapter 19

Logic Comparisons

DIGITAL

The logic we have been discussing in our previous chapters has used discrete on and off conditions for each element in a line of logic. In other words the logic depends upon the true or false status of each element. There are only two (2) states that an element can have, and no other variables. This is called *digital logic*. By combining a binary notation with digital logic, we can represent discrete numbers that can be used for encoding, decoding, counting, and various mathematical functions. Within a PLC or computer, we can also store ASCII characters to represent alpha-numeric characters that can be interpreted by a printer.

ANALOG

Another type of logic refers to continuous signals that represent a variable quantity over a continuous range. This is refered to as *analog logic*. Analog means "analogous to" or "equivalent to" another reference. Following are a few examples; a tachometer generates an analogous voltage signal in proportion to a shaft (mechanical) rpm, a thermistor controls current by changing resistance in proportion to changes in temperature; a pressure device controls a voltage or current in proportion to changes in pressure; a generator produces voltage in proportion to its excitation and speed, etc. Note that these are proportional values. The devices that change energy from one form to another are called *transducers.*

Analog signals, as a rule, are not linear over a very wide range, i.e., the ratio of input to output will not be constant due to changes in heat, resistance, magnetic saturation, etc. Therefore an analog system will tend to "drift" from its commanded performance over time.

There is also a condition known as *hysteresis* which is a tendency for a device or system to have two outputs for a given input. The output will tend to be low when the input rises, and remain high as the input drops. The *range* is selected over a relatively short spread of input changes so that it can be plotted as a relatively straight line and provide minimum hysteresis and distortion.

Variable processes are considered analog applications, and all variable outputs provide analog signals. Systems that have only analog controls are difficult to control with precision because of the drift and hysteresis. This is especially true when trying to synchronize multiple systems.

CONVERTERS

If we can combine the precision of digital logic with the variable analog logic, we can greatly improve process control. This requires interface devices to convert analog signals to and from digital formats.

A/D converter – This is the abbreviation for an analog to digital converter. This is a device or circuit that converts an analog signal to a digital signal. The digital signal is often represented as binary number that may provide a thousand (1000) or more discrete values across a range of perhaps 0 to 10 vdc.

D/A converter – This is the abbreviation for a digital to analog converter. The device or circuit converts digital signals to proportional analog signals. Again, the preciseness of the output depends upon how many discrete values the binary number has available. If the D/A converter has a number 0-1000 representing 0-10 vdc, a value of 500 would provide an output of 5 vdc. In other words this could produce a precise output with the accuracy of 0.01 vdc.

SIGNAL MANIPULATION

Multiple A/D converter signals, sensing various conditions, can be brought into our digital logic, probably by a PLC, and manipulated as necessary. For instance input temperature reference and feed-back signals might be sent to a *comparator* device or circuit which would in turn produce an *error* signal that could be used to correct any variance in system performance.

The various signals can be manipulated mathematically for internal scaling of values, adding, subtracting, multiplying, dividing, and displaying as digital read-outs to an operator's station. The final corrected calculations can then be sent out through a D/A converter as a precise reference voltage.

The advantage of this arrangement is that the continuous calculations and corrections minimize drift in the process. The digital signals always have discrete values that do not change until modified by a change in their bit pattern. In an all-analog system, we would likely be using resistors for voltage dividers and signal scaling. A pair of opposing generator fields might serve as our comparator to provide the appropriate error signals. This leaves the system susceptible to variable loading, resistance changes, and electro-magnetic hysteresis that would allow it to drift above and below the selected performance level.

Many industrial products have programmable digital *parameters* that can be used to adjust range limits and change signal scaling. This allows one to fine-tune variable logic for optimum performance. In some AC variable speed drive products there are as many as 2000 accessible parameters for adjustments and performance monitoring. Many power products have their own CPU on-board as well as A/D converters to accept different reference signals. They can also be linked to a PLC for digital remote control and process monitoring.

Chapter 20

Applied Analog Logic

In our previous chapter, we introduced the concept of analog logic. If you have to work with any process that has variables, you will likely have some analog logic to contend with. To provide an example of analog logic, we are going to look at an analog control system that was popular prior to the 1970's. It will illustrate ways the signals could be manipulated, and you will also see the many components that contributed to drift and hysteresis. This chapter will consider the analog logic required to control a DC motor.

DC DRIVE LOGIC

Industrial DC motors require a "drive system" to provide the necessary DC voltages to the motor. The drive system may use a solid state *Silicon Controlled Rectifier (SCR)* control, energized from an AC power source for the DC power, or a rotating DC generator that is driven by an AC motor. There are thousands of older installations in industry that still use a motor-generator set *(M-G)* for their power source. The common control system is often referred to as a *"Ward-Leonard"* control. This system connects the motor and generator armatures back to back so that the motor follows the voltage and polarity of the generator to establish its speed and direction of rotation. Additional auxiliary exciters (generators) are usually required to provide amplification and controlled excitation for the generator and motor fields. We will use a Ward-Leonard circuit to illustrate the DC motor logic because it allows us to follow the schematic without getting lost in the electronics of an SCR drive. The minimum logic we will discuss must be provided in all controls, electronic or otherwise.

HOW DOES A DC MOTOR WORK?

Before we attempt to define the logic required to control a DC motor, we need to review the way the DC motor is constructed, and how its operation is controlled. The mechanical/electrical construction of the DC motor is a lot more complex than an AC motor and requires special control considerations. Figure 20-1 illustrates a cutaway of a typical industrial DC motor.

There are two major parts to consider. One is the stator, or stationary "shell" of motor which holds the magnetic field poles. While some motors use permanent magnets, the vast majority are wound with a "shunt winding" to establish alternately North and South poles. This winding is separately excited to provide precise control of the field current. The available motor torque is determined by the flux density of this magnetic field as defined by the ampere-turns (amperes times coil turns) available. The stator will also have auxiliary poles between the shunt fields, called interpoles or commutating poles that carry armature current. Their purpose is to correct magnetic distortion, caused by the armature current and maintain a stable neutral zone for the brushes, to prevent sparking.

The rotating member, called the armature, has many coils wound in slots in the laminated iron core, and terminating at a current collector system called a commutator. The commutator consists of many insulated copper segments which collect current from the brush assembly, mounted in the end-bell of the motor.

When current is fed to the armature, magnetic poles are produced because of the unique position of the coils and brushes at any given time. The interaction between the field and armature magnetic fields produces torque as like poles repel, and opposite poles attract. As the armature rotates, the commutator continuously switches armature coils to maintain a constant magnetic relationship between the rotor and stator. The total available torque is a summation of the field and armature magnetic field strengths.

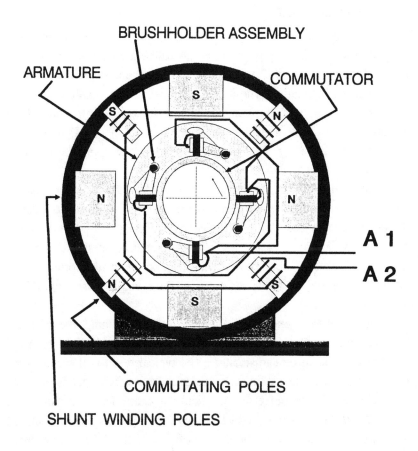

BRUSHHOLDER ASSEMBLY

ARMATURE

COMMUTATOR

A 1

A 2

COMMUTATING POLES

SHUNT WINDING POLES

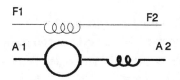

Figure 20-1 DC motor structure

SPEED REGULATION

DC motor armatures always generate a voltage which opposes the applied voltage. This opposing voltage is called a counter-electromotive force, abbreviated Cemf. The Cemf is dependent upon the field strength and the speed of rotation. The motor stabilizes its speed when the Cemf is theoretically equal to the armature source voltage. With constant field current, the armature speed varies directly with the applied voltage. When the armature and field both have full voltage applied, the motor will run at its *base speed.*

If the field strength is reduced while holding the armature voltage constant, the motor must run faster to generate enough Cemf to balance the input voltage. This is a useful characteristic that is used to drive the motor faster than the base speed. When operating in this mode, the motor has *constant horsepower* characteristics because torque (field strength) is falling off at the same rate as the speed is increasing. (Hp = Torque x Rpm/5252)

If the field loses excitation while the motor is running, the armature will accelerate until the residual magnetic field can produce the required Cemf before the speed can stabilize. The typical result is that the motor literally runs away, and ultimately explodes from centrifugal forces. For this reason, it is imperative that the field has controlled excitation at all times. A current-sensitive Field Loss relay or equivalent sensor should always be provided to take the system off line if the field current drops below a minimum level.

The Cemf is often used as a voltage feed-back in the control system to indicate that the motor has reached the commanded speed. The operator gives a reference command, and when Cemf matches it, the speed stabilizes at that point.

Cemf is also used to reverse current and provide counter-torque for controlled regenerative braking of high-inertia machines when generator voltage is reduced.

POWER SOURCES

Since there are two (2) separately controlled elements in a DC motor, we must supply two power sources. It is also imperative that field excitation is applied first, and kept at a safe operating level at all times. Armature voltage must be applied in a controlled manner, starting from zero, because current is only limited by resistance or Cemf. Armature resistances are very low – often less than 0.02 ohms. If high voltage is applied while at rest, the in-rush currents will cause extensive damage to the brush and commutator assemblies.

DC SCHEMATIC

We have established the "ground rules" for controlling the DC motor, and will now look at a relatively simple schematic showing a generator-driven DC motor and minimum control logic. See Figure 20-2.

As you scan down the center of the drawing, you can identify the power components that are being used. At the top on line 1, we show a single-phase, full-wave, bridge rectifier. We have not specified voltages, as they must be matched to the equipment you are working with. This rectifier is often referred to as the *constant potential* source. In many installations, a constant-potential *exciter* (generator) may be installed. Many older installations now have rectifiers that have replaced a failed exciter.

The motor field exciter *MFX*, is shown as a generator having two control windings, *AF1-AF2*, and *AF3-AF4*, with its output connected directly across the motor field terminals *MF1-MF2*, through a current-sensing "field-loss" relay. The exciter could be a small generator called an *amplidyne*, once manufactured by General Electric Co. A few milliamperes of current on a control field can produce up to 250 vdc and 6 amperes output for the motor field excitation. This amplification allows use of small, low current devices to control very large power circuits.

The generator and motor armatures are shown connected back-to-back to form the "power loop". This circuit has an instantaneous overload relay to protect against sudden stalling loads, and an isolating loop contactor *M*. The contactor must be rated to carry the expected DC current, and will require DC arc blow-out features. The logic is designed to prevent opening this contactor under load, but it should be able to handle any emergency service at rated current.

At line 11, we have the generator field exciter *GFX*, which is identical in construction to *MFX*. If the generator field required a totally different current capacity, a different type of exciter could be used.

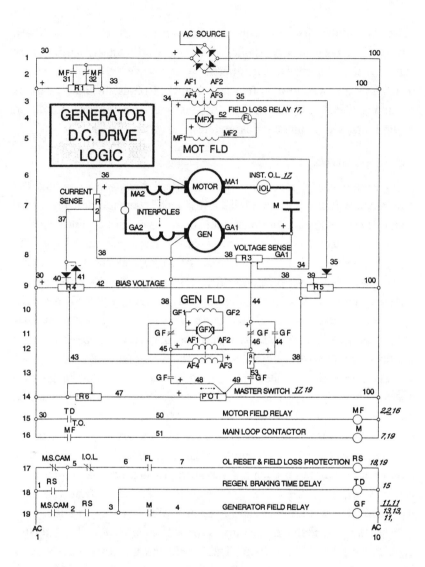

Figure 20-2 Generator, DC drive logic

CONTROL COMPONENTS

Our relay logic is shown at the bottom of the schematic. All relays could be DC, or the DC *MF* and *M* could have special AC coils. We have shown both AC and DC to identify components that usually have DC rated contacts. When amplidynes are used, the excitation currents are so low that AC relays that can break the low energy DC currents, are sometimes applied as *GF* and *MF* relays.

The placement of the *M* and *MF* in the DC control circuit is a carry-over from the time when a rotating constant potential exciter was used instead of the bridge rectifier. In that case, if AC power was lost, the inertia of the rotating components could provide DC control power to hold in *M* and *MF* long enough to bring the motor to a controlled regenerative stop.

The *master switch (controller)* is constructed as a reversible "throttle" assembly. When at neutral position (off), brushes #48 and #49 are both contacting the negative side of the constant potential power – with no difference in potential. The potentiometer is constructed in such a way that #48 picks up increasing positive voltage and #49 stays on the nega-tive bus in the forward direction. When operating in reverse, #49 picks up increasing positive voltage while #48 stays on the negative bus. The master switch also has a pair of cam-operated switches shown of lines 17 and 19 in the neutral, off position.

The high side AC control voltage will require a control On/Off switch (not-shown) ahead of wire #1.

OPERATING SEQUENCE

We start with all control power on, including the bridge rectifier. The motor field exciter *MFX* will receive a minimum excitation on its field *AF1-AF2* through resistor tap #32 on resistor *R1*. The resistor tap should be adjusted so that the output will provide minimum rated motor field current and pull-in the field failure relay *(FL)*.

If the *Master Switch* is in the off position, *RS* will now be energized and seal-in. If there is an overload, or a field loss, *RS* will drop out immediately. *RS* closes an enabling contact on line #19 as long as it remains energized. If an instantaneous overload trips *RS* off, returning the master switch to *OFF* will reset it. This insures that the system must return to zero speed, after any fault, before it can restart.

As the operator moves his master switch in either direction, the cam at #1-#2 closes to instantly energize the off-delay timer *TD*. *TD* immediately picks up the motor field relay *MF* which switches the motor field excitation to maximum on *R1*, line 2. This applies full field on the motor for maximum torque.

MF on line 16 closes and energizes *M* to close the power loop between the armatures. *M* on line 19 then enables *GF* to connect the reference field of the generator field exciter *GFX* to the master switch circuits #48-#49. *GFX* amplifies this signal and energizes the generator field at *GF1-GF2*. As the generator excitation builds up, current flows in the power loop and the motor accelerates as it follows the voltage. Instantaneous polarities are shown on the schematic for this condition.

VOLTAGE REGULATION

Generator output voltage is reflected across resistor *R3,* wires #38 and GA1. Tap #44 senses a percentage of generator voltage across the resistor to wire #38 with #44 having the most positive voltage. We can use this adjustable voltage as a feed-back signal to our reference. You will see that the reference signal to the generator field exciter passes through resistor *R7* and at this time #46 is more positive than #53. The excitation current creates a voltage drop across the resistor. If we can run another current in the same direction through the same resistor, we can increase its voltage drop, (increase apparent resistance) and reduce the current that is flowing through the field *AF1-AF2*. By adding a positive tap #44 on R7 nearest to #46, and a negative tap #38 nearer to #53, we can provide an adjustable potential that is proportional to the generator voltage, and use it as a feed-back to stabilize the voltage (speed) when the feed-back approximates the original reference.

CURRENT REGULATION

Resistor *R2* is connected across the motor and generator interpole coils using interpole taps #36 and #38. A current surge will create an inductive voltage boost across the interpoles which appears across *R7*. This is often found to be about 7 volts on many installations with about

200% of rated loop current. We can calibrate this signal by adjusting tap #37 with reference to #38. At this instant, #37 will be most positive.

R4 and *R5* are called bias resistors, and are connected across the constant potential bus. When #30 is positive, the current will pass through the diode at #40, through the resistor to #43, and then through the *GFX* field *AF4-AF3* to #38. #30 is the positive end of R4, and #40 is more positive than #43, so we again can have two currents passing in the same direction through a resistor. If we adjust the standing difference in potential between #40 and #43 to 5 volts, no control current will flow until the potential across #37 and #38 exceeds 5 volts. If the signal reaches 7 volts, then the bias circuit would allow the equivalence of 2 volts to pass across *R4* and through the field *AF4-AF3*. Excitation is in opposite polarity to the reference, so this field will subtract from the reference signals command. Thus when there is a surge of current, this serves as a current limit, and over-rides the command reference to protect our system. We can adjust the intensity of the signal on *R2* and set its threshold on *R4* for the point where it is to become effective. The adjustment between #41 and #43 sets the bias voltage for operation in the opposite direction.

FIELD CROSSOVER

A wood veneer lathe drive often operates in the constant horsepower range because it must speed up to maintain veneer sheet speed as the log gets smaller in diameter. Reverse is used to back out of a jam, so does not require the field weakening control. The circuit shown is quite primitive for 1995, but the concept is valid. In this case we are using the *MFX* field *AF4-AF3* to subtract from the reference field, and in turn weaken the motor field for higher speeds. The voltage signal is derived from tap #34 on R3, and wire #38. #34 is most positive and we use taps #38 and #39 on R5 to set the "spill over" armature voltage that we want to reach before we start field weakening. Note that #39 must be set more positive than #38. Tap #34 on *R3* determines the maximum signal that we need to pick off.

REGENERATIVE BRAKING

If we suddenly return our "throttle" to zero, *GF* will immediately drop out, and transfer the reference field of *GFX* to the voltage appearing across #38 and #34. Note that this voltage always has a polarity that is opposite to the reference voltage. *TD* will hold *MF* and *M* energized for a preset interval, usually about 3 seconds.

While the motor is still turning, with full field excitation, it is generating considerable voltage. Since this voltage opposes the generator output, the current in the power loop reverses, but if you look closely the voltage across #38-GA1, it does not change polarity. This voltage is applied through *GF* to the *GFX* in reverse polarity and drives the Generator output to zero. This is often called a "suicide circuit." The motor comes to a stop at a rate controlled by the current limit circuit, whose polarity has also reversed to cushion the stop. If there is no command to restart within about 3 seconds, *MF* will be released by *TD*, reducing the motor field to a stand-by condition and then deenergize *M* to open the power loop.

When the master switch is operated in opposite direction, all polarities reverse, and action is the same, except in this instance, field weakening will not be available.

COMMENTS

There has been no attempt in this discussion to "engineer" a drive system because it would be necessary to calculate all resistor values and wattages to match the requirements of all the control fields. We would need to know the resistance of all machine fields, their current and voltage ratings and the amplification ratios of the exciters. Some additional resistors might be required to provide proper scaling of the feed-back signals.

Figure 20-2 is a generic "demonstration" circuit to show how the logic might be applied. A system, using this circuit, would be susceptible to considerable "drift" attributable to temperature changes.

In a modern system, most of the feed-back signals would be scaled and passed directly through A/D converters, processed digitally, then returned as reference signals through a D/A converter. The digital numbers could be manipulated mathematically so that none of the bias resistors would be required. The motor and generator field exciters could be replaced with silicone controlled rectifier (SCR) power supplies, and controlled directly from the digital logic.

If the generator is replaced by a "power SCR" module, it would be controlled directly, eliminating the generator field exciter. When using digital control on all solid-state components, very few analog devices will be required so very precise control of the DC motor can be accomplished. The essential requirements for controlling the motor can be provided by either analog or digital logic – or a combination of both. The arrival of solid-state components has provided a means of doing it most efficiently.

Chapter 21

Unusual Trouble-shooting Case Histories

Since our stated purpose in this book is to help electricians and technicians become better trouble-shooters, it is appropriate to include some interesting real-life experiences that illustrate a few challenges we have experienced. Often the apparent problem could be quickly identified, but the real fault was hard to "pin down."

RELAY SYSTEMS

Many elevator and marine control panels have been constructed in the past as live-front boards. This construction uses an asbestos-composition or "Glasstex" insulating board for its principal structure. Contactors and relays are assembled directly on this board which might be 3/4" to 1-1/4" thick. Most wiring is installed on the back of board and connected to brass or copper bolts that secure various parts of the respective components. Stationary contacts are bolted directly to the board. Movable contacts will have a flexible, braided shunt fastened to a through-bolt. Power and control voltages are usually DC.

With a good schematic, the circuitry is usually fairly easy to follow. Wire numbers and component identifications are frequently stamped on the front and back of the board. Since the panel is usually not enclosed, it is subject to any contamination that might be in the area, such as salt air, diesel fumes, oil-mist, carbon dust from brushes, etc.

Board Cracks

With aging, some boards tend to "craze" or form hair-line cracks between mounting bolts. This fault may collect conductive contaminates over time. The resistance of the contaminates may be high enough that you might not detect it with an ohmmeter, but when voltage is applied, there is enough leakage current to cause a fault. Correction of this problem depends upon whether or not the crack can be routered-out enough to clear the conducting material and replace it with an insulating compound. It may be necessary to bore a hole in another area and move the wiring away from the defect. Components must be isolated and tested with a "megger" (high-voltage resistance tester) to prove adequate insulation. Ohmmeters usually operate on relatively low-voltage (1.5-9 vdc) while a megger uses about 500 vdc.

Shunts

The flexible shunt strands will crack from fatigue over time. If they are carrying much current, they will develop hot spots, and may even char the board. In time they will break. Since the shape has become conformed to the terminal locations, it may not be visually apparent that it is broken.

Through-bolts

One of the most frustrating problems is to trace a faulty circuit across the back of the board, then trace it on the front of the board without finding any open circuit. In many cases there is a line-up of boards, making it difficult for one person to check both sides simultaneously. The conducting through-bolts are made of relatively soft metal (brass or copper). If they have been over-stressed during assembly, they may eventually break within the board, and still support any attachments. The only way to find this open circuit is to test for continuity from both ends of the bolt.

Contacts

On one occasion, a plywood mill in central Oregon called our plant in Seattle, WA requesting immediate service as their DC veneer lathe was shut-down, and total production was in jeopardy. The author was flown to the site to correct their problem. The circuit was very similar to Figure 20-2, using a live-front DC panel board.

It was immediately obvious that maximum motor field was not being applied so that the motor would not start. Every relay and contactor was operating in proper sequence. The motor field relay was energized, and its contacts were closed, but the motor field voltage did not respond. We connected a voltmeter across the closed contacts and read full DC voltage, indicating an open circuit.

The silver-plated contacts were removed for inspection. We found a glassy crystalline material had formed on the contact faces, which was totally non-conductive. We scraped the deposit off, touched the face up with a little silver solder, and had the mill back in service in about 20 minutes. Apparently, the silver plating had reacted with the moisture and contaminates in the mill basement to form a non-conducting salt that was invisible to the casual observer.

LIGHTNING

Residential

Following a severe thunderstorm, we received a call from a lady whose electric range had quit. Upon arrival, we found that one of the burners had burned out, but there were all sorts of other strange things that had happened to her electrical system. While replacing fuses in her entrance panel, we noticed that the bare ground wire had scorched the side of her house where it led to the ground rod and cold water pipe.

Behind her house was a small travel trailer that had been finished with a fiberboard siding and trimmed with metal flashing. A length of "romex" wiring cable had been stretched from the house to the trailer for temporary power. The outer insulation on the cable had burst at about one foot intervals from one end to the other. The metal flashing had separated and bent outward at every splice, and a small radio was literally "cremated" inside the trailer. A small can of kerosene had burned briefly beneath her sink. During the following year we continued to find "tracks" of that lightning strike in various homes in that area in the form of charred spots and premature appliance failures.

The lightning apparently had struck the distribution transformer and used this lady's home for a lightning rod. The point is that this sort of phenomenon can alert one to continuing patterns to watch for if abnormal faults seem to occur.

Municipality

We received a call from a municipal water district to check out the pump at the city well. The pump had blown a fuse, and could not be restarted after replacement. This was a new installation, so we checked it out thoroughly, finding that there was a ground and an open circuit in the wiring to the submersible pump.

The pump had to be removed from the deep well at considerable expense. When brought out of the well, they found that the wiring had been twisted around the pipe and broken, and the pump was unscrewed, to within the last couple of turns, from the pipe. In fact, they almost lost the pump during retrieval. As in most political affairs there was a lot of finger pointing by all concerned, including the local utility. We did determine that there had been a lightning storm at about the time of the failure.

Submersible pumps are usually threaded to the bottom of the riser pipe. Since the pipe has a right-hand thread, the pump is screwed on in a clockwise direction. The pump rotates in a counter-clockwise direction so that starting torque tends to tighten the threaded connection.

The 230 volt, three phase motor was apparently running when the lightning struck the utility lines. The lightning strike produced DC current with voltage in the "megavolt" range. If you apply DC current to a running induction motor, it will try to "freeze" the rotor. This technique is often applied for braking. In this case, the DC current stopped the motor rotor, which was running about 3500 rpm, and the inertia, in the counter clockwise direction unscrewed the pump and tore off the leads. There was no fault to be shared.

HARMONICS

In recent years, there have been increasing problems with "neutral" conductors in commercial and industrial installations. The neutral conductor typically is expected to carry the unbalanced current in a grounded-Y or single phase 120/240 volt feeder system. This usually allowed us to install a neutral conductor, one size smaller than the load conductors.

With the growth of electronics in all types of electrical installations, we now have a new problem known as *harmonics*. All electronic devices control power by taking timed bursts of current from the 60 Hz power source. If the current bursts are large enough, they will reflect disruptive secondary frequencies of current and voltage back to the power line.

Computers have now become a predominate load in commercial buildings, and their switched power supplies reflect a 3rd harmonic back into the distribution system. This 180 Hz energy is not cancelled out in a 60 Hz system, and becomes an added load on the neutral conductor. In some instances the neutral current becomes greater than the phase current. If we have an old installation, with small neutral conductors, harmonic currents can create very hazardous heat conditions.

As an example, a few years ago, the Seattle Opera House upgraded its rheostat dimming system to a solid-state control. The neutral system was almost "burned down" before maintenance people discovered this unexpected harmonics problem.

The new electric codes are now addressing this problem, and often require the neutral conductor to be considerably larger than phase conductors. If you suddenly find neutral conductor problems, one should consider if the load characteristics have changed.

A failed neutral can cause havoc in an installation where we have 120/240 single phase service. Without the neutral conductor, 240 volts will be divided across the loads depending upon the relative impedance of the loads connected to each "hot" line. Some loads may only have 60 volts across them while others have 180 volts. Obviously motors may stall, or lamps burn out under these conditions.

GROUND FAULTS

Fire pump

A sawmill had a fire pump installed on a creek that was across the road from the plant site. They complained that fuses were frequently blowing out on this important installation. Finally it got to the point where the fuse would blow every time it was replaced. The pump was supplied power from overhead weatherproof conductors which were fed from a mill shed.

We isolated the pump and found that it was clear of any faults. Back at the service panel, we noted that the system ground wire had burned a track on the wall, indicating that it had been subjected to high currents. We knew at this time that we had a power ground fault somewhere in our system, and it had to be in the feeder lines to the pump.

At first, there did not appear to be any visible problems as we walked the right-of-way, inspecting the overhead lines. Finally, as we circled each pole for closer inspection, we found one of the weatherproof power wires was touching a guy-wire. We climbed the pole and discovered that the wind had been swinging the wire against the guy-wire until it had worn through the insulation. This was causing intermittent fuse failures whenever the wind was blowing. In the final instance the wire had welded it self to the guy-wire, making a permanent ground path. The ground resistance in that area was low enough to pass enough current to blow fuses.

Blower

The same mill site had a planer mill built across the yard from the sawmill. A 100 Hp blower was used to blow the chips and waste from the planer to a waste burner near the sawmill. (This was before environmental concerns had outlawed such burners.) One morning, as the operator turned on the blower during start-up, he felt an electrical shock, heard a loud noise and saw a puff of smoke from the blower motor. The operator was one-armed, and was using his "hook" to operate the reduced voltage starter for the blower. Needless to say, he was a pretty excited operator by the time we arrived on the scene.

We climbed up into the overhead to inspect the motor, and found that the cast-iron motor junction box had a one inch hole blown through its cover. Inside, we found that taping on the bolted connections to one phase had worn through and made contact with the junction box cover. It was a simple procedure to rework the connection and re-insulate it. The mill had to restart, so the motor was put back on line, and work continued.

Since this fault was on a 460 vac power circuit, we knew that there had to be another grounded power lead somewhere in the plant. There was no ground fault detection equipment so we would have to search for the fault, circuit by circuit. Often a single grounded power lead will not be evident until a second fault occurs. We had to wait for a weekend shut-down to begin our testing.

We checked every circuit in the planer mill and found nothing amiss. The sawmill was several hundred yards away, on the other side of the yard. We searched through its circuitry for considerable time, and finally found that a feeder to the carriage drive M-G set had a grounded lead. It turned out that the fault was in an "LB" conduit fitting where a cover screw had pierced the insulation of one of the conductors. The problem was corrected, and no other faults were found.

The interesting thing about this problem was the fact that the faults were in apparently unrelated parts of the mill site. The common connection between the two sites was the waste burner. A bank of draft fans was installed at the burner, with power derived from the sawmill system. We found "tracks" of the fault in conduit connections to motor starters at these fans, where concentric knock-out holes were blown out around the conduit pipes. The current apparently followed the metal work from the fans to the burner shell, and then back to the planer mill through its blower pipe.

PHANTOM FAULT

Did you ever determine that you had a fault, when there really wasn't one? This experience is almost an embarrassment, but may save someone else from the same experience.

In the early 1950's we were wiring a new retirement house on the bank of the Rogue River in southern Oregon. The owner decided that he wanted a switch by his bed that he could turn on in the morning, and start his coffee pot in the kitchen, about 30 "wire feet" away. That should be simple enough! Just run an extended switch circuit in series with his coffee outlet. This was done.

The owner was wandering around with a little neon test lamp, and was "supervising" our work as we turned power on the various circuits. He decided to test his "coffee circuit" and came running to us complaining that the bedroom switch did not turn off the kitchen outlet. Using his neon tester, we found that he was right! Using a voltmeter we got a reading of about 10 volts. Do we have an insulation problem?

Convinced that we had a defective cable, we proceeded to remove it and install new wiring. He tested our circuit again, and found that we had the same condition.

At that point, we decided that we were missing something in our analysis. We took a new roll of 2-wire cable, and connected the black wire to 120 volts, and left the white wire disconnected. We took a meter reading between the two wires and found about 10 volts AC. The voltage was produced by the mutual inductance between the two wires. The 60 Hz field around the energized wire was inducing a small voltage on the wire laying in parallel with it. The neon lamp could not tell the difference between 10 volts and 120 volts. (As we were picking up our tools, I noticed that the neon tester was laying under the house, and appeared to have an accidental footprint of one my electricians on it.)

Some of these incidents are laughable as we reflect upon them, but were not so funny as they did cost real time and material at the time. They have been included to point out that things are not necessarily as they seem.

Chapter 22

Conclusion

"Fundamentals of Electrical Control" is a subject that seems to have been neglected by most writers. Over the years, this author has searched for a down-to-earth text to recommend to aspiring apprentices and electricians who are interested in more than installing wires. As mentioned in the Foreword, most publications are written to people who already have considerable experience. Most basic information is directed to a specific product or process that limits general application.

We have attempted to help one understand <u>how</u> to approach any type of logic so that he is not limited by pre-conceived ideas. <u>You may not decide to design logic, but you do need the skills to interpret what others have designed, to be a successful trouble-shooter.</u>

As this text has evolved, it is easy to see how difficult it is to produce an <u>objective</u> presentation. The materials were first presented to a group of over twenty elevator maintenance electricians, with experience ranging from 3 months to 20+ years. We deliberately did not use elevator circuits for primary instruction to break their "mind-set" from elevator installations, and get them thinking. When a complex elevator schematic was introduced later in their course, it was exciting to watch everyone, even the least experienced, "jump" into the discussions as each rung was interpreted.

In our earlier chapters, we showed that *logic* is relative to everything we do. If we remember the premise that we can use <u>known conditions to predict and control</u> future events we can apply logic. Remember

that we even applied a ladder diagram to an abstract problem. If we
have trouble reaching a logical conclusion, we just drop back and determine what the pre-requisites are, solve them and move forward.

The discussion of components in Chapter 2 was expanded to
include application considerations. When trouble-shooting, it is important to recognize a mis-applied device that may be causing faults. A
failed component should not be replaced by an improper device that
could create greater problems. This is especially important when working with DC systems.

After looking at some ladder logic examples, we worked through
the steps required to design a simple control system. This exercise
showed one how to think through sequential logic, step by step, illustrating the interdependence of related rungs of logic. This was followed
by developing working drawings required to assemble the actual project.
The start-up chapter simulated a real-life experience, complete with
some "glitches".

The "tilt-hoist" example led you through a system that relied on
"position" for control logic. The infeed conveyor used position inputs to
provide a simple, unattended, automatic operation. It is not unusual to
find multiple, position related, rungs of logic that move materials
between machines on a factory floor. The materials move through each
area, looking ahead for its next position, while also resetting a previous
position to receive a following product. With a little ingenuity, one can
expand these principles to build a large complex system.

By participating in the design of ladder diagrams, one should get a
"feel" for how the designer thinks. What are the required steps? What
are the pit-falls that must be avoided? Do I need to add logic to protect
against a fault? Can I change my logic or inputs to eliminate a trouble-
some line of logic? All these questions indicate some of the decisions
that one may have to make.

A parenthetical chapter was inserted to show several unique logic
schemes that use relays for special functions. Versions of these circuits
are used throughout industry for moving, sorting, and counting products.
If you understand how these work, it makes it easier to use similar logic
in programmable logic controllers.

Chapter 12 introduced you to the programmable logic controller
(PLC). Editors were concerned that this subject might be so complex

that readers would be "turned off" by too much "computerese". While we did use an Allen-Bradley PLC-5 system as our model, our intent was to show how relay ladder logic relates to PLC programs. All systems require similar hardware, and have very similar instruction sets. The numbering systems are common to many PLC's and computers, so they are applicable for many applications.

Documentation and display of PLC programs is also very similar to the relay logic, and can be interpreted with reasonable ease, once the addressing systems are understood. Trouble-shooting can often be done at the CRT terminal, without having to run all over a plant floor.

In Chapter 17, we converted our original conveyor logic to a PLC program for comparison. While there were some unique changes in the hard-wiring, you can follow the original logic through the software program.

Up to this point, all the logic was "digital", in that it used the ON/OFF condition of discrete inputs and outputs for all control. No variable parameters were used at any time. In Chapter 19 we introduced "analog" logic which uses continuously variable signals. A combination of analog and digital logic is used to control variable processes.

The DC motor control system has both "digital" relays and variable "analog" components, so it was used to illustrate the contrasting logic. You can see where the analog signals are continuously variable, and proportional signals are derived for system control. Relays controlled the ON/OFF functions.

This step by step presentation of logic, from basic digital systems through PLCs and analog systems, tried to illustrate *fundamentals* without getting bogged down in too much subjective material. We hope that our reader can relate to the principles illustrated by the examples, and apply them to his day to day requirements.

Chapter 21 provided a number of case histories covering some unusual trouble-shooting incidents. While you may not encounter any of these conditions, it does show the strange turn of events that sometimes turn up when trouble-shooting.

As a parting remark, "It can really be can be fun, making a living, from other peoples problems!" A good trouble-shooter is always in demand. Happy trouble-shooting!

Appendix A

Glossary

Addressing A system of labeling that identifies I/O terminals with memory locations in a PLC.

ASCII Abbreviation for American Standard Code for Information Interchange, and is used to transmit data between computers and peripheral equipment. Each character requires one byte of information.

A/D converter A device that converts an analog signal to a digital signal

Amplification Magnification of a variable such as voltage, current, or power.

Analog Refers to a signal representing a variable quantity over a continuous range. A proportional signal.

Armature The rotating member of a DC motor or generator.

Automatic process Defines a process that can operate in a predictable manner, using available inputs, without continual intervention of a human operator.

BCD Binary coded decimal. Applying binary numbers to portray decimal numbers, usually for read-out purposes. See Chapter 13.

Binary   A number system with a base of 2. Binary digits are used to represent the on/off states of devices numerically as 0 = off, 1 = on. Used for notation in digital control systems. See Chapter 13.

Bit   The smallest increment of digital information. Has only two conditions, on (1) and off (0).

Boolean logic   A mathematical representation of logic, named after the English logician George Boole.

Bridge rectifier A full wave rectifier for converting current from AC to DC.

Brush   Usually a carbon conductor used to maintain electrical connection between rotating and stationary parts of a machine. e.g. a DC motor.

Byte   Eight (8) contiguous bits of information. Half of a word. The minimum increment of information that can transmit an ASCII character.

Cemf   Counter electromotive force. The product of a motor armature rotating in a magnetic field, always opposing the applied voltage.

Coding   A process of converting information or signals according to a set of rules with finite numbers or symbols. Language used in computer or logic circuits.

Commissioning   The task of inspecting and successfully putting a process into production. Start-up services.

Commutator   A cylindrical assembly of insulated conducting segments, used on DC armatures to transfer current through brushes to the armature windings. Windings terminate at commutator segments in a pattern that form discrete magnetic poles in the armature.

Comparator	Device that compares one signal to another, such a process signal to a reference.
Control	Signal used to direct, supervise, or manage power.
Controller	Device that controls power. May include other elements such as a comparator.
Control transducer	Device that changes a low-level signal to another form of low-level signal, such as converting a temperature signal to an electrical signal.
Converter	A device that changes AC current to DC current.
D/A converter	Device or circuit that converts a digital signal to an analog signal.
Decimal	A numbering system with a base of ten (10). See Chapter 13.
Decode	The act of retrieving encoded information for designated operations.
Digital	Information that is expressed by signals that rely on discrete states, such as on and off. See Binary.
Diode	Device that passes current in one direction and blocks current in the opposite direction.
Drift	A slow change in output over time with constant input and load, usually related to component temperature changes.
EEPROM	Electrically erasable read only memory.
Encode	Store selected identification data in numerical format.
Encoder	A device or circuit that converts information by means of a code.
Error	Difference between a reference signal and a process signal, usually used to make system corrections.

Energy	The ability to do work. Energy and work can be expressed in the same units. Machines transform or transmit energy.
Feedback	A part of the output signal that is used to compare with input references to determine error signal.
Field control	A method of controlling DC motor speed by adjusting the magnitude of the current in shunt field windings.
File	A group of words in a PLC. Files are labelled according to their special functions.
Flow chart	A method of diagramming sequential logic, identifying actions to be performed, with branching alternatives. It is often used as an outline to guide more detailed logic development.
Force	A push or pull that tends to change the position or motion of an object. Usually measured in ounces or pounds.
Frequency	Number of periodic cycles per unit of time.
Hexadecimal	A numbering system with a base of sixteen (16). See Chapter 13.
Histogram	A histogram displays the transition history of a data table value on the CRT screen. Data can be saved to disk and printed.
Hysteresis	Property of a system that causes the output to be two-valued for a given change in input. Output tends to be low with a rising change, and high for a lowering change. Commonly refers to the magnetic behavior of ferromagnetic materials.
Input	A term usually applied to command or status information used to direct a process. This can be digital or analog information.

Instruction	Programmable command used to develop PLC logic.
Inverter	A device that changes DC current to AC current.
I/O	Designates input/output devices used in PLC applications.
Jog	Momentary control of an operation, sometimes called "inching". Used principally on motor-driven machines for set-up, emergency operations, where continuous operation may be hazardous.
Kinetic energy	The energy of motion possessed by a body
Ladder logic	A method of representing logic in a format resembling a ladder. Each rung on the ladder represents a single line of logic, defining conditions to produce an action or store status information.
Logic	The science of reasoning according to a specified set of rules. See text Chapter 1.
Manual process	An operation that requires "hands-on" control of a human operator.
NEC	National Electric Code. Electrical recommendations by the National Fire Protection Association.
NEMA	National Electrical Manufacturers Association, a non-profit organization that has established standard specifications for electrical equipment.
Octal	A numbering system with a base of eight (8). See Chapter 13.
Output	A term applied to devices that produce motion or alter conditions as a result of logical commands.
Parallel	A circuit where one or more commands will produce the same end result. Often called an "OR" function in logic.

Permissive logic   Refers to "safety circuits" that must be satisfied before a process can operate. This logic can usually pre-empt all other controls and stop a process, or prevent it from starting.

PID   Proportional, integral, derivative process control algorithm required to fine tune such quantities as temperature, pressure, flow rates, fluid levels an precise speed controls.

Potential energy   The energy possessed by a body when at rest.

Power   Work done per unit of time. Force exerted over a distance. 1 hp = 33000 ft-lb/min = 746 watts.

Power transducer   Device that changes power from one form to another, such as an electrical motor that converts electrical to mechanical power.

Program   A set of instructions or operations arranged so that a machine can perform a series of operations.

PLC   A programmable logic controller for machines or processes where the sequence of operations can be changed easily by programming. The central processing unit (CPU) contains logic circuitry, memory and interface to input-output devices.

Rectifier   A device that transforms that alternating current to direct current.

Regenerative braking   Form of braking where the kinetic energy of a motor and load is returned the system power supply.

ROM   Read-only memory

Saturation   Condition of a system where a further increase in input no longer results in an appreciable change in output.

Scan	The period of time a PLC takes to read the input image table, perform the logic, write to the output image table, and activate outputs, and return to start.
Schematic	A drawing that illustrates logic in a sequential manner, without respect to physical location of components.
Sensor	Control transducer that gathers information and converts it to a signal.
Sequencer	Another name for a Central Processing Unit (CPU).
Series	Circuit arrangement where the same current passes through each element. In logic, a circuit where two or more commands are required in sequence to produce the same end result. An "AND" function in logic.
Set point	The commanded reference level at which a load or process should operate.
Shift register	A logic circuit which can move encoded information through multiple zones, where the codes can be read, and selectively acted upon.
Signal	Medium that carries information (electrical, mechanical, fluid). Information at low power level and a function of time.
Software	A term applied to programmable information. The process logic in a PLC.
Torque	The turning effort applied to a shaft to cause rotation. Measured in ounce-inches or pound-feet.
Truth table	A table comparing input conditions with resulting outputs for logic functions. See Figure 15-2.
Transducer	Device that changes energy from one form to another such as speed to voltage signal.

Ward Leonard   Method of controlling DC motor speed by control-
                     ling armature voltage.

Wiring diagram   A drawing that shows the physical point-to-point
                     wiring of an installation. Wiring diagram is
                     "keyed" to its schematic by proper annotation.
                     Wire and conduit sizing, and terminal locations are
                     usually a part of a wiring diagram.

Word   Sixteen (16) bits, or two bytes of information.

Work   A force moving an object over a distance. Meas-
                     ured in inch-ounces or foot-pounds. Work = force x
                     distance.

Appendix B

Formulas

HORSEPOWER FORMULAS

For Rotating Objects

$$HP = \frac{T \times N}{5252}$$

Where T = Torque (lbft)
N = Speed (rpm)

Objects in Linear Motion

$$HP = \frac{F \times V}{33000}$$

Where F = Force (lbs)
V = Velocity (ft/min)

For Pumps

$$HP = \frac{GPM \times Head \times Specific\ Gravity}{3960 \times Efficiency\ of\ Pump}$$

$$HP = \frac{GPM \times PSI \times Specific\ Gravity}{1713 \times Efficiency\ of\ Pump}$$

Where GPM = Gallons per Minute
Head = Height of Water (ft)
Efficiency of Pump = %/100
PSI = Pounds per SqIn

Specific Gravity of Water = 1.0
1 CuFt per Sec. = 448 GPM
1 PSI = A Head of 2.309 ft (water weight)
62.36 lbs per CuFt at 62ßF

For Fans and Blowers

$$HP = \frac{CFM \times PSF}{33000 \times \text{Efficiency of Fan}}$$

$$HP = \frac{CFM \times PIW}{6356 \times \text{Efficiency of Fan}}$$

$$HP = \frac{CFM \times PSI}{229 \times \text{Efficiency of Fan}}$$

Where CFM = Cubic Ft per Minute
PSF = Pounds per SqFt
PIW = Inches of Water Gauge
PSI = Pounds per SqIn
Efficiency of Fan = % / 100

For Conveyors

$$HP(\text{verticle}) = \frac{F \times V}{33000}$$

$$HP(\text{horizontal}) = \frac{F \times V \times \text{Coef. of Friction}}{33000}$$

Where F = Force (lbs)
V = Velocity (ft/min)

Coef. of Friction
Ball or Roller Slide = .02
Dovetail Slide = .20
Hydrostatic Ways = .01
Rectangle Ways with Gib = 0.1 to 0.25

TORQUE FORMULAS

$$T = \frac{HP \times 5252}{N}$$

Where T = Torque (LbFt)
HP = Horsepower
N = Speed (Rpm)

$$T = F \times R$$

Where T = Torque (LbFt)
F = Force (Lbs)
R = Radius (Ft)

$$Ta \text{ (accelerating)} = \frac{WK^2 \times \text{Change inRPM}}{308 \times t(\text{Sec})}$$

Where Ta = Torque (LbFt)
WK^2 = Inertia at Motor Shaft (LbFt)2
t = Time to Accelerate (Sec)

Note: To change $LbFt^2$ to $InLbSec^2$, Divide by 2.68,
To change $InLbSec^2$ to $LbFt^2$, Mult. by 2.68.

AC MOTOR FORMULAS

$$\text{Sync Speed} = \frac{\text{Freq} \times 120}{\text{Number of Poles}}$$

Where Sync Speed = Synchronous Speed (Rpm)
Freq = Frequency (Hz)

$$\% \text{ Slip} = \frac{(\text{Sync Speed - FL Speed}) \times 100}{\text{Synch Speed}}$$

Where FL Speed = Full Load Speed (Rpm)
Sync Speed = Synchronous Speed (Rpm)

$$\text{Reflected } WK^2 = \frac{WK^2 \text{ of Load}}{(\text{Reduction Ratio})^2}$$

ELECTRICAL FORMULAS

Ohms Law

$$I = \frac{E}{R}, \quad R = \frac{E}{I}, \quad E = I \times R$$

Where I = Current (Amperes)
    E = EMF (Volts)
    R = Resistance (Ohms)

Power in DC Circuits

$$P = I \times E \qquad HP = \frac{I \times E}{746}$$

$$kW = \frac{I \times E}{1000}, \qquad kWH = \frac{I \times E \times Hours}{1000}$$

Where P = Power (Watts)
    I = Current (Amperes)
  kW = Kilowatts
  WH = Kilowatt Hours

Power in AC Circuits

$$kVA\ (1ph) = \frac{I \times E}{1000} \qquad kVA\ (3ph) = \frac{I \times E \times 1.73}{1000}$$

Where kVA = Kilovolt Amperes
     I = Current (Amperes)
    E = EMF (Volts)

$$kW\ (1ph) = \frac{I \times E \times PF}{1000}$$

$$kW\ (2ph) = \frac{I \times E \times PF \times 1.42}{1000}$$

$$kW\ (3ph) = \frac{I \times E \times PF \times 1.73}{1000}$$

$$PF = \frac{W}{V \times I} = \frac{kW}{VA}$$

Where kW = Kilowatts
    I = Current (Amperes)
    E = EMF (Volts)
    F = Power Factor
    W = Watts
    V = Volts
  kVA = Kilovolt Amperes

Calculating Motor Amperes

$$\text{Motor Amperes} = \frac{HP \times 746}{E \times 1.732 \times Eff \times PF}$$

$$\text{Motor Amperes} = \frac{kVA \times 1000}{1.73 \times E}$$

$$\text{Motor Amperes} = \frac{kW \times 1000}{1.73 \times E \times PF}$$

Where HP = Horsepower
    E = EMF (volts)
  Eff = Efficiency (% / 100)
  kVA = Kilovolt Amperes
  kW = Kilowatts
    F = Power Factor

Calculating AC Motor Locked Rotor Amperes

$$LRA = \frac{HP \times \left(\dfrac{Start\ kVA}{HP}\right) \times 1000}{E \times 1.73}$$

Where LRA = Locked Rotor Amperes
    HP = Horsepower
    VA = Kilovolt Amperes
    E = Volts

$$LRA\ @\ Freq\ (x) = \frac{60\ Hz\ LRA}{\left(\dfrac{60}{Freq(x)}\right)^{1/2}}$$

Where 60 Hz LRA = Locked Rotor Amperes
Freq (x) = Desired Frequency (Hz)

OTHER FORMULAS

Calculating Accelerating Force for Linear Motion

$$\text{Fa (Acceleration)} = \frac{W \times \text{Change in V}}{1933 \times t}$$

Where Fa = Force (Lbs) of Acceleration
   W = Weight (Lbs)
   V = Velocity (Fpm)
   t = Time to Accel. Wt. in Seconds

Calculating Min. Accelerating Time of a Drive

$$t = \frac{WK^2 \times N \text{ change}}{308 \times Ta}$$

Where t = Time to Accel. Load in Seconds
   WK^2 = Total Inertia (LbFT^2)
   N = Speed (Rpm)
   Ta = Accelerating Torque (LbFt)

Note: To change LbFT^2 to InLbSec^2, Divide by 2.68.
   To change InLbSec^2 to LbFt^2, Multiply by 2.68.

$$RPM = \frac{FPM}{262 \times D}$$

Where RPM = Revolutions per Minute
   FPM = Feet per Minute
   D = Diameter (Feet)

$$WK2 \text{ reflected to Motor} = \text{Load WK2} \times \left(\frac{\text{Load RPM}}{\text{Motor RPM}}\right)^2$$

Where WK2 = Inertia (LbFt)2
   RPM = Revolutions per minute

Index